# TEST BANK

# Introduction to the Human Body
## The Essentials of Anatomy and Physiology

Fifth Edition

**Gerald J. Tortora**
Bergen Community College

**Sandra Reynolds Grabowski**
Purdue University

## Prepared by

**Karin VanMeter**
Des Moines Area Community College

JOHN WILEY & SONS, INC.
NEW YORK • CHICHESTER • WEINHEIM • BRISBANE • SINGAPORE • TORONTO

COVER PHOTO: ©Al Satterwhite//FPG International.

To order books or for customer service call 1-800-CALL-WILEY (225-5945).

QS 50

A

P

ISBN 0-471-38327-9

Printed in the United States of America

10  9  8  7  6  5  4  3  2  1

Printed and bound by Victor Graphics, Inc.

**Table of Contents**

# Chapter 1    Organization of the Human Body

## Multiple-Choice

*Choose the one alternative that best completes the statement or answers the question.*

1. The science dealing with the functions of the body parts is called
   A) physiology.
   B) cytology.
   C) anatomy.
   D) histology.
   E) biology.
   Answer: A

2. The level of organization when different tissues join together is called the
   A) chemical level.
   B) cellular level.
   C) tissue level.
   D) organ level.
   E) system level.
   Answer: D

3. Group of related organs that have a common function is called a(n)
   A) organ.
   B) system.
   C) tissue.
   D) group.
   E) organism.
   Answer: B

4. The process by which unspecialized cells become specialized cells is called
   A) anabolism.
   B) catabolism.
   C) metabolism.
   D) differentiation.
   E) homeostasis.
   Answer: D

5. The sum of all chemical processes in our body is
   A) anabolism.
   B) catabolism.
   C) metabolism.
   D) differentiation.
   E) homeostasis.
   Answer: C

6. All of the following are examples of organs *except:*
   A) stomach.
   B) heart.
   C) muscle.
   D) brain.
   E) gallbladder.
   Answer: C

7. Hair and nails belong to the
   A) skeletal system.
   B) excretory system.
   C) immune system.
   D) integumentary system.
   E) endocrine system.
   Answer: D

8. The condition in which the body's internal environment stays within physiological limits is
   A) responsiveness.
   B) homeostasis.
   C) differentiation.
   D) growth.
   E) All of the above are correct.
   Answer: B

9. All of the following belong to feedback systems which control homeostasis EXCEPT
   A) control center.
   B) receptor.
   C) receiver.
   D) effector.
   E) All of the above are basic components of the feedback system.
   Answer: C

10. In a negative feedback system, the response of the effector
    A) enhances the original stimulus.
    B) eliminates the original stimulus.
    C) reverses the original stimulus.
    D) does not change the original stimulus.
    E) does not affect the original stimulus.
    Answer: C

11. Reproduction refers to:
    A) increase in the number of cells.
    B) the formation of new cells.
    C) production of a new individual.
    D) Both B and C are correct.
    E) Both A and B are correct.
    Answer: D

12. A structure which sends inputs to the control center is a(n)
    A) effector.
    B) receptor.
    C) affector.
    D) stimulus.
    E) output structure.
    Answer: B

13. A baroreceptors measures
    A) touch.
    B) blood pressure.
    C) temperature.
    D) alveolar stretch.
    E) pain.
    Answer: B

14. Which of the following is a symptom of disease rather than a sign?
    A) nausea
    B) bleeding
    C) vomiting
    D) fever
    E) rash
    Answer: A

15. The science that deals with the drug treatment of disease is called
    A) physiology.
    B) anatomy.
    C) epidemiology.
    D) pharmacology.
    E) pharmacy.
    Answer: D

16. In the anatomical position the subject
    A) is lying down.
    B) has arms placed above the head.
    C) has arms folded on the chest.
    D) is standing upright facing the observer with the palms backwards.
    E) is standing upright facing the observer with the palms forward.
    Answer: E

17. The plane that divides the body into unequal left and right portions is the
    A) parasaggital plane.
    B) midsaggital plane.
    C) frontal plane.
    D) transverse plane.
    E) oblique plane.
    Answer: A

18. A gluteal injection refers to an injection into the
    A) buttock.
    B) lower limb.
    C) ankle.
    D) upper limb.
    E) trunk.
    Answer: A

19. The sternum is _____ to the heart.
    A) posterior
    B) anterior
    C) inferior
    D) lateral
    E) distal
    Answer: B

20. The plane that divides the body into a superior and inferior portion is the
    A) parasaggital plane.
    B) midsaggital plane.
    C) transverse plane.
    D) oblique plane.
    E) frontal plane.
    Answer: C

21. Which of the following is NOT considered a basic tissue type?
    A)  connective tissue
    B)  epithelial tissue
    C)  cartilage tissue
    D)  nervous tissue
    E)  muscle tissue
    Answer:  C

22. Which of the following organs does NOT belong to the digestive system?
    A)  liver
    B)  gallbladder
    C)  ureter
    D)  stomach
    E)  salivary glands
    Answer:  C

23. The organ system which regulates the body's activities through hormones is the
    A)  digestive system.
    B)  endocrine system.
    C)  nervous system.
    D)  cardiovascular system.
    E)  integumentary system.
    Answer:  B

24. The _____ is the region between the lungs from the breastbone to the vertebra.
    A)  vertebral canal
    B)  pericardium
    C)  mediastinum
    D)  pleural cavity
    E)  manubrium
    Answer:  C

25. The organ system that transports fats from the gastrointestinal tract to the cardiovascular system is the
    A)  digestive system.
    B)  endocrine system.
    C)  lymphatic system.
    D)  urinary system.
    E)  respiratory system.
    Answer:  C

26. The anatomical term which best describes a structure toward the head is
    A)  superficial.
    B)  deep.
    C)  inferior.
    D)  superior.
    E)  anterior.
    Answer:  D

27. The ring finger is _____ to the little and middle fingers
    A)  lateral
    B)  intermediate
    C)  deep
    D)  distal
    E)  proximal
    Answer:  B

28. The anatomical term for navel is
    A) crural.
    B) inguinal.
    C) umbilical.
    D) femoral.
    E) coxal.
    Answer: C

29. The best anatomical term to describe the back region of the body would be
    A) ventral.
    B) dorsal.
    C) gluteal.
    D) vertebral.
    E) popliteal.
    Answer: B

30. The brain and the spinal cord are located in the
    A) ventral cavity.
    B) dorsal cavity.
    C) abdominal cavity.
    D) thoracic cavity.
    E) vertebral cavity.
    Answer: B

31. All of the following belong to the ventral body cavity EXCEPT
    A) thoracic cavity.
    B) abdominal cavity.
    C) cranial cavity.
    D) pleural cavity.
    E) pericardial cavity.
    Answer: C

32. The body cavity containing the urinary bladder and portions of the large intestine is the
    A) pelvic cavity.
    B) abdominal cavity.
    C) mediastinum.
    D) pleural cavity.
    E) dorsal cavity.
    Answer: A

33. The pericardial cavity contains the
    A) lungs.
    B) thyroid glands.
    C) brain.
    D) heart.
    E) stomach.
    Answer: D

34. Which of the following does NOT belong to the nine-abdominopelvic regions?
    A) left lumbar region
    B) right upper quadrant
    C) right iliac region
    D) epigastric region
    E) hypogastric region
    Answer: B

35. The majority of the stomach is found in the
    A) right hypochondriac region.
    B) left hypochondriac region.
    C) epigastric region.
    D) hypogastric region.
    E) umbilical region.
    Answer: C

36. Failure of the body to maintain homeostasis will
    A) have no effect on health.
    B) cause illness.
    C) always causes death.
    D) initiate positive feedback.
    E) increase body temperature.
    Answer: B

37. All of the following are controlled by homeostatic mechanisms, EXCEPT
    A) muscle movement.
    B) blood pressure.
    C) body temperature.
    D) blood sugar levels.
    E) breathing rate.
    Answer: A

38. Which of the following processes is controlled by positive feedback?
    A) blood sugar levels
    B) blood clotting
    C) blood pressure
    D) body temperature
    E) breathing rate
    Answer: B

39. Collectively the organs are called
    A) mediastinum.
    B) mammary glands.
    C) abdominal organs.
    D) viscera.
    E) pleural organs.
    Answer: D

40. All of the following are functions of the cardiovascular system, EXCEPT
    A) oxygen transport.
    B) carbon dioxide transport.
    C) red blood cell production.
    D) blood clot formation.
    E) transport of hormones.
    Answer: C

41. The basic structural and functional units of an organism are
    A) atoms.
    B) molecules.
    C) cells.
    D) tissues.
    E) organs.
    Answer: C

42. Atoms combine to form
    A) cells.
    B) organs.
    C) organ systems.
    D) molecules.
    E) None of the above are correct.
    Answer: D

43. The organ system which contains the skin and its derivatives is the
    A) skeletal system.
    B) integumentary system.
    C) muscular system.
    D) endocrine system.
    E) cardiovascular system.
    Answer: B

44. The anatomical term used to describe the region of the neck is
    A) facial.
    B) brachial.
    C) cervical.
    D) thoracic.
    E) mental.
    Answer: C

45. The splitting of proteins into amino acids is an example of
    A) anabolism.
    B) catabolism.
    C) metabolism.
    D) differentiation.
    E) homeostasis.
    Answer: B

## True-False

*Write T if the statement is true and F if the statement is false.*

1. Anatomy refers to the structure and function of the body.
   Answer: False

2. The chemical level of structural organization includes all chemicals needed to maintain life.
   Answer: True

3. The integumentary system protects all body systems.
   Answer: True

4. Anabolism is a chemical process, which does not require energy.
   Answer: False

5. Reproduction refers to the formation of new cells for growth only.
   Answer: False

6. Stress has an adverse effect on homeostasis, because it creates an imbalance in the internal environment.
   Answer: True

7. Even with abnormally low blood pressure the brain will still receive adequate amounts of oxygen.
   Answer: False

8. The control center determines the level at which the controlled condition needs to be maintained.
   Answer: True

9. To describe the relationship of different body structures to each other, anatomists use directional terms.
   Answer: True

10. The sagittal plane divides the body into equal superior and inferior portions.
    Answer: False

11. The spinal cord and the brain are located in the cranial cavity.
    Answer: False

12. The heart is located in the mediastinum.
    Answer: True

13. For practical purposes clinicians prefer to use the quadrant division of the abdominopelvic region.
    Answer: True

14. Childbirth is a good example of a positive feedback mechanism.
    Answer: True

15. The heart is superior to the cranium.
    Answer: False

## Short Answer

*Write the word or phrase that best completes each statement or answers the question.*

1. The study dealing with the structures of the human body is called _____.
   Answer: anatomy

2. All the body systems combined make up a(n) _____.
   Answer: organism

3. Molecules combine to form the _____ level of organization.
   Answer: cellular

4. The highest level of organization is the _____.
   Answer: organismic level

5. The ability to detect and respond to changes in the environment is called _____.
   Answer: responsiveness

6. The maintenance of relatively stable conditions for the cells of the human body is called _____.
   Answer: homeostasis

7. The process by which unspecialized cells become specialized cells is called _____.
   Answer: differentiation

8. _____ is the force of blood as it passes through the arteries.
   Answer: blood pressure

9. The basic component of a feedback system that produces a response is the _____.
   Answer: effector

10. The component of a feedback system that monitors changes in the controlled condition is the _____.
    Answer: receptor

11. The skin participates in the production of vitamin _____.
    Answer: D

12. The plane that divides the body into anterior and posterior positions is the _____ plane.
    Answer: frontal

13. The fluid surrounding body cells is called _____ fluid.
    Answer: interstitial

14. The stomach is _____ to the lungs.
    Answer: inferior

15. The molecules produced and released by the endocrine system are referred to as _____.
    Answer: hormones

16. The humerus is _____ to the radius
    Answer: proximal

17. The _____ cavity contains the heart.
    Answer: pericardial

18. The thoracic and abdominopelvic cavities are subdivisions of the _____ body cavity.
    Answer: ventral

19. The correct anatomical term for the back of the elbow is _____.
    Answer: olecranon

20. The anatomical term for the head is _____.
    Answer: cephalic

21. All the bones of the body, their associated cartilages, and joints belong to the _____ system.
    Answer: skeletal

22. Hair and nails belong to the _____ system.
    Answer: integumentary

23. Blood clotting is an example of _____ feedback.
    Answer: positive

24. Homeostasis is mainly controlled by the _____ and _____ systems.
    Answer: nervous and endocrine

25. The basic structural and functional units of an organism are _____.
    Answer: cells

## Matching

*Choose the item from Column 2 that best matches each item in Column 1.*

*Match each term in Column 2 with its definition in Column 1*

1. Column 1: The study of the structure of the human body
   Column 2: Anatomy

2. Column 1: The study of the function of the human body
   Column 2: Physiology

3. Column 1: The lowest level of structural organization
   Column 2: Chemical level

4. Column 1: The sum total of all chemical reactions in the human body.
   Column 2: Metabolism

5. Column 1: A chemical reaction that uses energy.
   Column 2: Anabolism

*Choose the correct anatomical term from Column 2 for each of the common names in Column 1.*

6. Column 1: eye
   Column 2: orbital

7. Column 1: hand
   Column 2: manual

8. Column 1: cheek
   Column 2: buccal

9. Column 1: buttock
   Column 2: gluteal

10. Column 1: spinal column
    Column 2: vertebral

## Essay

*Write your answer in the space provided or on a separate sheet of paper.*

1. Name and briefly describe the levels of structural organization in the human body.

   Answer: The chemical level: includes atoms and molecules.
   The cellular level: includes all different cells made of combinations of molecules.
   The tissue level: tissues consist of groups of similar cells.
   The organ level: organs are formed when different types of tissues join together.
   The system level: consists of related organs that have a common function.
   The organismic level: the highest level of structural organization includes all organ systems making up the entire organism.

2. Name and briefly describe the planes that can be passed through the human body.

   Answer: A sagittal plane divides the body into right and left portions.
   A midsagittal plane divides the body into equal right and left portions.
   A parasagittal plane divides the body into unequal right and left portions.
   A frontal (coronal) plane divides the body into anterior and posterior portions.
   A transverse plane divides the body into superior and inferior portions.
   An oblique plane passes through the body in an angle between the planes mentioned above.

3. Define homeostasis.
   Answer: Homeostasis is a condition in which the internal environment of the body is maintained within certain physiological limits.

## CHAPTER 2   Introductory Chemistry

## Multiple-Choice

*Choose the one alternative that best completes the statement or answers the question.*

1. Negatively charged particles in an atom are called
   A) neutrons.
   B) electrons.
   C) protons.
   D) elements.
   E) isotopes.
   Answer: B

2. When two or more atoms combine, the resulting combination is called
   A) atomic mass.
   B) atomic number.
   C) inert element.
   D) molecule.
   E) salt.
   Answer: D

3. The chemical symbol for sodium is
   A) K.
   B) O.
   C) Mg.
   D) Na.
   E) Sn.
   Answer: D

4. Substances that cannot be broken down into simpler substances by ordinary chemical reactions are called
   A) matter.
   B) compounds.
   C) chemical elements.
   D) inorganic molecules.
   E) organic molecules.
   Answer: C

5. The number of protons in an atom corresponds with the
   A) atomic mass.
   B) atomic number.
   C) number of neutrons and electrons.
   D) atomic weight.
   E) number of valences.
   Answer: B

6. An abundant element found in water and most organic molecules is
   A) nitrogen.
   B) hydrogen.
   C) potassium.
   D) carbon.
   E) sodium.
   Answer: B

7. The backbone of organic molecules is formed by
   A) carbon.
   B) oxygen.
   C) magnesium.
   D) phosphorus.
   E) nitrogen.
   Answer: A

8. A particle with a negative or positive charge is referred to as
   A) electron.
   B) neutron.
   C) ion.
   D) proton.
   E) isotope.
   Answer: C

9. The atomic number of oxygen is
   A) 4.
   B) 6.
   C) 7.
   D) 8.
   E) 12.
   Answer: D

10. A chemical bond where electrons are transferred from one atom to another is a(n)
    A) hydrogen bond.
    B) ionic bond.
    C) single covalent bond.
    D) double covalent bond.
    E) polar covalent bond.
    Answer: B

11. An atom that gives up electrons is considered to be a(n)
    A) electron acceptor.
    B) electron donor.
    C) anion.
    D) electron.
    E) molecule.
    Answer: B

12. The bond which is found between water molecules is a(n)
    A) hydrogen bond.
    B) ionic bond.
    C) single covalent bond.
    D) double covalent bond.
    E) None of the above are correct.
    Answer: A

13. A chemical bond in which one pair of electrons is shared between atoms is a(n)
    A) hydrogen bond.
    B) ionic bond.
    C) single covalent bond.
    D) double covalent bond.
    E) triple covalent bond.
    Answer: C

14. The most common chemical bonds in the human body are
    A) covalent bonds.
    B) ionic bonds.
    C) hydrogen bonds.
    D) double bonds.
    E) None of the above.
    Answer: A

15. Which of the following best describes a synthesis reaction?
    A) Large molecules are broken down to form smaller ones.
    B) Molecules combine to form large molecules.
    C) A reaction that always requires chemical energy.
    D) A and C are correct.
    E) B and C are correct.
    Answer: E

16. All of the following are organic compounds EXCEPT
    A) nucleic acids.
    B) water.
    C) proteins.
    D) lipids.
    E) carbohydrates.
    Answer: B

17. Energy needed for chemical reactions in the body is provided by the breakdown of
    A) ribonucleic acid (RNA).
    B) deoxyribonucleic acid (DNA).
    C) adenosine diphosphate (ADP).
    D) adenosine triphosphate (ATP).
    E) adenosine monophosphate (AMP).
    Answer: D

18. In an average healthy adult, 55% to 60% of the body weight is composed of
    A) amino acids.
    B) salts.
    C) water.
    D) fat.
    E) proteins.
    Answer: C

19. All of the following are properties of water EXCEPT
    A) it is an excellent solvent.
    B) it absorbs heat very quickly.
    C) it can participate in chemical reactions.
    D) it serves as a lubricant.
    E) it releases heat very slowly.
    Answer: B

20. A molecule that dissociates in water and gives off hydrogen ions is a(n)
    A) salt.
    B) acid.
    C) base.
    D) buffer.
    E) solvent.
    Answer: B

21. The higher the number on the pH scale,
    A) the higher the hydrogen ion concentration.
    B) the higher the hydroxide ion concentration.
    C) the more acidic a solution.
    D) the more neutral a solution.
    E) the lower the hydroxide ion concentration.
    Answer: B

22. Which of the following describes the most acidic solution?
    A) pH 4
    B) pH 5
    C) pH 7
    D) pH 9
    E) pH 14
    Answer: A

23. The pH of blood is
    A) 4-5.
    B) 6.76-7.00.
    C) 7.20-7.60.
    D) 7.35-7.45.
    E) 7.65-8.00.
    Answer: D

24. To prevent drastic changes in the pH and to maintain homeostasis, the body
    A) uses digestive enzymes.
    B) increases the hydrogen ion concentration in the blood.
    C) decreases the hydrogen ion concentration in the blood.
    D) uses buffer systems.
    E) changes the body temperature.
    Answer: D

25. The storage form of glucose in the liver is
    A) fructose.
    B) glycogen.
    C) fat.
    D) starch.
    E) glycerol.
    Answer: B

26. The building blocks of carbohydrates are
    A) polysaccharides.
    B) disaccharide.
    C) monosaccharides.
    D) glycogen.
    E) starches.
    Answer: C

27. In humans, glycogen is stored in the cells of the
    A) brain.
    B) liver.
    C) muscles.
    D) Both A and B are correct.
    E) Both B and C are correct.
    Answer: E

28. Polysaccharides can be broken down into simpler sugars by the process of
    A) dehydration synthesis.
    B) simple synthesis.
    C) hydrolysis.
    D) anabolism.
    E) None of the above are correct.
    Answer: C

29. The most highly concentrated source of energy in the body is
    A) proteins.
    B) amino acids.
    C) glycogen.
    D) triglycerides.
    E) glucose.
    Answer: D

30. The building blocks of triglycerides are
    A) glycerol and fatty acids.
    B) chains of fatty acids.
    C) monosaccharides.
    D) chains of amino acids.
    E) nucleic acids.
    Answer: A

31. Chemically, certain sex hormones such as estrogens and testosterone are classified as
    A) proteins.
    B) carbohydrates.
    C) nucleic acids.
    D) lipids.
    E) starches.
    Answer: D

32. All of the following belong to the group of unsaturated fats EXCEPT
    A) palm oil.
    B) olive oil.
    C) sunflower oil.
    D) canola oil.
    E) corn oil.
    Answer: A

33. Chemically estradiol is a(n)
    A) amino acid.
    B) steroid.
    C) protein.
    D) enzyme.
    E) nucleic acid.
    Answer: B

34. The building blocks of proteins are
    A) carbons.
    B) nucleic acids.
    C) amino acids.
    D) glycerol.
    E) fatty acids.
    Answer: C

35. When three amino acids combine, the result is a(n)
    A) dipeptide.
    B) tripeptide.
    C) octapeptide.
    D) nonapeptide.
    E) polypeptide.
    Answer: B

36. The bonds formed between amino acids are
    A) peptide bonds.
    B) ionic bonds.
    C) hydrogen bonds.
    D) nitrogen bonds.
    E) None of the above occur between amino acids.
    Answer: A

37. All of the following statements about enzymes are true EXCEPT
    A) they speed up chemical reactions without changing themselves.
    B) enzymes are proteins.
    C) they are highly specific.
    D) they are biological catalysts.
    E) they are used up in a chemical reaction.
    Answer: E

38. All of the following are enzymes EXCEPT
    A) lactose.
    B) peptidase.
    C) oxidase.
    D) aminopeptidase.
    E) amylase.
    Answer: A

39. The building blocks of nucleic acids are
    A) amino acids.
    B) fatty acids.
    C) ribonucleic acids.
    D) nucleotides.
    E) peptides.
    Answer: D

40. Which of the following molecules make up the genetic information in a cell?
    A) ATP
    B) RNA
    C) DNA
    D) ADP
    E) mRNA
    Answer: C

41. Which of the following molecules contains the sugar ribose?
    A) RNA
    B) DNA
    C) ATP
    D) ADP
    E) AMP
    Answer: A

42. Which of the following molecules is considered to be a double helix?
    A) mRNA
    B) tRNA
    C) RNA
    D) DNA
    E) ATP
    Answer: D

43. Energy required to make ATP is supplied by a process called cellular
    A) transcription.
    B) dehydration.
    C) translation.
    D) respiration.
    E) reproduction.
    Answer: D

44. Which of the following is NOT a base found in DNA?
    A) Guanine
    B) Thymine
    C) Uracil
    D) Cytosine
    E) Adenine
    Answer: C

45. The double helix of DNA is stabilized by _____ bonds.
    A) peptide bonds
    B) hydrogen bonds
    C) ionic bonds
    D) polar covalent bonds
    E) none of the above
    Answer: B

## True-False

*Write T if the statement is true and F if the statement is false.*

1.  Chemical elements present in high concentrations in the human body are trace elements.
    Answer: False

2.  An element is composed of the same type of atoms.
    Answer: True

3.  Each electron shell of an atom can hold six electrons.
    Answer: False

4.  A compound is a chemical composed of two or more different elements.
    Answer: True

5.  Positively charged ions are called anions.
    Answer: False

6.  A covalent bond is more stable than a hydrogen bond.
    Answer: True

7. Decomposition reactions break down chemical bonds.
   Answer: True

8. Inorganic compounds usually contain carbon.
   Answer: False

9. The combination of a solvent and a solution is called a solute.
   Answer: False

10. Water requires a large amount of heat to change from a liquid to a gas.
    Answer: True

11. A solution with a pH of 7 is slightly basic.
    Answer: False

12. Carbohydrates are composed of carbon, hydrogen, and oxygen.
    Answer: True

13. Glycogen is a monosaccharide.
    Answer: False

14. Cholesterol is used to produce some sex hormones.
    Answer: True

15. A change in the shape of a protein is called denaturation.
    Answer: True

## Short Answer

*Write the word or phrase that best completes each statement or answers the question.*

1. Anything that occupies space and has mass is considered to be ____.
   Answer: matter

2. The portion of an atom which contains protons and neutrons is the ____.
   Answer: nucleus

3. The total number protons and neutrons in an atom is its ____.
   Answer: mass number

4. Atoms, which have a completely filled outer electron shell, are called _____.
   Answer: inert elements

5. Positively charged atoms are called ____.
   Answer: cations

6. An electrically charged ion or molecule that has an unpaired electron in its outermost shell is a(n)
   _____.
   Answer: free radical

7. Substances in the cell that can combine with free radicals are called _____.
   Answer: antioxidants

8. Atoms that pick up electrons from other atoms are called _____.
   Answer: electron acceptors

9. In covalent bonds, the electrons of two atoms are _____.
   Answer: shared

10. A chemical reaction in which a molecule is split apart is referred to as a(n) _____
    reaction.
    Answer: decomposition

11. Synthesis reactions are chemical reactions that _____ energy.
    Answer: require

12. The capacity to do work is _____.
    Answer: energy

13. A substance that can be dissolved in a solvent is called a _____.
    Answer: solute

14. Another term for "water loving" is _____.
    Answer: hydrophilic

15. When salts dissolve in water they undergo _____.
    Answer: ionization (dissociation)

16. To describe the acidity or alkalinity of a solution, the _____ scale can be used.
    Answer: pH

17. The higher the hydrogen ion concentration, the more _____ the solution.
    Answer: acidic

18. Monosaccharides and disaccharides are referred to as _____ sugars.
    Answer: simple

19. When many monosaccharides are joined together through dehydration synthesis a
    _____ is formed.
    Answer: polysaccharide

20. A fat in which all the carbon atoms are bonded to hydrogen atoms is called _____ fat.
    Answer: saturated

21. When two amino acids join together they form a _____.
    Answer: dipeptide

22. The portion of an enzyme where the substrate binds is called the _____.
    Answer: active site

23. Substances that speed up chemical reactions but do not change themselves are called

    _____.
    Answer: catalysts (enzymes)

24. The nitrogen base in DNA that always pairs with adenine is _____.
    Answer: thymine

25. When a phosphate group is removed from ATP, the molecule _____ is formed.
    Answer: ADP

## Matching

*Choose the item from Column 2 that best matches each item in Column 1.*

*Match each term in Column 2 with its definition in Column 1.*

1.   Column 1: Negatively charged particles of an atom.
     Column 2: Electron

2.   Column 1: The smallest unit of matter.
     Column 2: Element

3.   Column 1: A negatively charged ion.
     Column 2: Anion

4.   Column 1: The number of deficient electrons in the outermost shell.
     Column 2: Valence

*Match the chemicals in Column 1 with the group of organic compounds in Column 2.*

5.   Column 1: glycogen
     Column 2: carbohydrate

6.   Column 1: steroids
     Column 2: lipid

7.   Column 1: enzyme
     Column 2: protein

8.   Column 1: RNA
     Column 2: nucleic acid

9.   Column 1: prostaglandin
     Column 2: lipid

10.  Column 1: fructose
     Column 2.: carbohydrate

## Essay

*Write your answer in the space provided or on a separate sheet of paper.*

1.   Name the different chemical bonds and briefly describe how they are formed.
     Answer:  Covalent bonds are formed by the sharing of one, two or three pairs of
             electrons.
             Ionic bonds are formed when an actual transfer of electrons occurs between
             atoms.
             Hydrogen bonds are the weakest of the chemical bonds and are formed due
             to an attraction of a hydrogen atom of one molecule and an oxygen or
             nitrogen atom of another molecule.

2.   Briefly explain the importance of water in the human body.
     Answer: Water makes up about 55 to 60 percent of body weight and therefore is the most abundant
             substance in the body. Water is an excellent solvent and suspending medium, it absorbs and
             releases heat very slowly, helping to maintain the homeostasis of body temperature.

3. Explain the difference between acids, bases, and salts.
   Answer: Acids are compounds, which dissociate in water and give off hydrogen ions. Bases give off hydroxide ions when they dissociate in water, and salts ionize in water into cations and anions, neither of which are hydrogen or hydroxide ions.

## CHAPTER 3   Cells

## Multiple-Choice

*Choose the one alternative that best completes the statement or answers the question.*

1.  The plasma membrane consists of
    A)  cellulose and carbohydrates.
    B)  proteins mostly.
    C)  entirely of phospholipids.
    D)  phospholipids, proteins, and carbohydrates.
    E)  carbohydrates and lipids.
    Answer: D

2.  The cytoplasm is the term for
    A)  all cell organelles combined.
    B)  microtubules and microfilaments.
    C)  the fluid portion of the cell.
    D)  the cytosol plus cell organelles.
    E)  the communication center of the cell.
    Answer: D

3.  Which of the following statements are true for the plasma membrane?
    A)  It is selectively permeable.
    B)  It contains glycoproteins.
    C)  It contains cholesterol.
    D)  A and B are correct.
    E)  A, B, and C are correct.
    Answer: E

4.  The lipid bilayer is permeable to all of the following substances EXCEPT
    A)  amino acids.
    B)  fat-soluble vitamins.
    C)  steroids.
    D)  oxygen.
    E)  water.
    Answer: A

5.  Endocytosis is an example of
    A)  excretion.
    B)  passive transport.
    C)  active transport.
    D)  facilitated diffusion.
    E)  simple diffusion.
    Answer: C

6.  The movement of solvents and dissolved substances across a selectively permeable membrane
    by gravity it is called
    A)  osmosis.
    B)  facilitated diffusion.
    C)  simple diffusion.
    D)  filtration.
    E)  active transport.
    Answer: D

7. Which of the following is necessary for diffusion to take place?
   A) a concentration gradient
   B) a selectively permeable membrane
   C) a hypertonic solution
   D) cellular energy
   E) All of the above.
   Answer: A

8. A red blood cell placed in a hypotonic solution
   A) loses water.
   B) gains water.
   C) neither gains nor loses water.
   D) shrinks.
   E) will not change shape.
   Answer: B

9. An isotonic solution for human red blood cells is a
   A) 10 % NaCl solution.
   B) 2% saline solution.
   C) 0.9% saline solution.
   D) water.
   E) 0.9% glucose solution.
   Answer: C

10. Pinocytosis and phagocytosis are a function of the
    A) cytoplasm.
    B) plasma membrane.
    C) ribosomes.
    D) mitochondria.
    E) cell nucleus.
    Answer: B

11. The movement of molecules from area of low concentration to an area of high concentration requires
    A) cellular energy.
    B) facilitated diffusion.
    C) integral proteins.
    D) Both A and B.
    E) Both A and C.
    Answer: E

12. Certain white blood cells can destroy bacteria by the process of
    A) pinocytosis.
    B) phagocytosis.
    C) exocytosis.
    D) lysis.
    E) None of the above.
    Answer: B

13. Which of the following statements BEST describes active transport?
    A) a concentration gradient is needed
    B) requires a carrier
    C) requires cellular energy
    D) requires osmotic pressure
    E) both B and C
    Answer: E

14. Chromatin is found in the
    A) nucleus.
    B) nuclear pores.
    C) ribosomes.
    D) mitochondria.
    E) lysosomes.
    Answer: A

15. The cell organelles which are continuous with the nuclear envelope are the
    A) lysosomes.
    B) Golgi complexes.
    C) endoplasmic reticulum.
    D) mitochondria.
    E) centrosomes.
    Answer: C

16. The packaging and sorting of proteins is the function of the
    A) endoplasmic reticulum.
    B) Golgi complex.
    C) mitochondria.
    D) lysosomes.
    E) nucleus.
    Answer: B

17. Steroid synthesis is the function of the
    A) Golgi complex.
    B) ribosomes.
    C) rough endoplasmic reticulum.
    D) smooth endoplasmic reticulum.
    E) mitochondria.
    Answer: D

18. Digestive enzymes are found in
    A) Golgi complexes.
    B) rough endoplasmic reticulum.
    C) smooth endoplasmic reticulum.
    D) lysosomes.
    E) mitochondria.
    Answer: D

19. Protein synthesis occurs at the
    A) smooth endoplasmic reticulum.
    B) rough endoplasmic reticulum.
    C) mitochondria.
    D) Golgi complexes.
    E) lysosomes.
    Answer: B

20. Which of the following are considered the "powerhouses" of the cell?
    A) lysosomes
    B) ribosomes
    C) nucleoli
    D) mitochondria
    E) peroxisomes
    Answer: D

21. Which of the following belongs to the cytoskeleton?
    A) cytosol
    B) mitochondria
    C) microtubules
    D) centromere
    E) flagella
    Answer: C

22. The structure located near the nucleus, made of two cylindrical structures composed of clusters of microtubules is the
    A) nucleolus.
    B) centrosome.
    C) flagellum.
    D) microtubule.
    E) cilium.
    Answer: B

23. Which of the following cell organelles help to detoxify the blood in the liver?
    A) nucleus
    B) lysosomes
    C) rough ER
    D) smooth ER
    E) vacuoles
    Answer: D

24. The process by which worn-out cell organelles are digested is called
    A) autolysis.
    B) autoregulation.
    C) autophagy.
    D) lysis.
    E) phagocytosis.
    Answer: C

25. The human somatic cells contain _____ chromosomes.
    A) 46
    B) 23
    C) 43
    D) 24
    E) 54
    Answer: A

26. Which of the following is a component of RNA only?
    A) andenine
    B) cytosine
    C) guanine
    D) thymine
    E) uracil
    Answer: E

27. The anticodon is located on the
    A) mRNA.
    B) rRNA.
    C) tRNA.
    D) DNA.
    E) ribosome.
    Answer: C

28. The process of translation during protein synthesis takes place
   A) in the nucleus.
   B) in the nucleolus.
   C) in the cytoplasm.
   D) on the ribosomes.
   E) on the mitochondria.
   Answer: D

29. Amino acids that participate in protein synthesis are picked up in the cytosol by
   A) mRNA.
   B) rRNA.
   C) tRNA.
   D) ribosomes.
   E) ATP.
   Answer: C

30. A CGT base triplet on DNA is copied into mRNA as
   A) CAT.
   B) GCA.
   C) GCU.
   D) ACU.
   E) TGC.
   Answer: B

31. When particular protein is complete, synthesis is stopped by a special
   A) anticodon.
   B) amino acid.
   C) stop codon.
   D) start codon.
   E) carrier protein.
   Answer: C

32. All of the following are nucleotide bases in DNA molecules EXCEPT
   A) adenine.
   B) cytosine.
   C) guanine.
   D) thymine.
   E) uracil.
   Answer: E

33. A protein is defined by the sequence of its
   A) fatty acids.
   B) amino acids.
   C) molecules.
   D) atoms.
   E) enzymes.
   Answer: B

34. The cell division which produces two identical cells is called
   A) somatic cell division.
   B) meiosis I.
   C) meiosis II.
   D) cytokinesis.
   E) reproductive cell division.
   Answer: A

35. The cytoplasmic division is referred to as
    A) meiosis.
    B) somatic cell division.
    C) reproductive cell division.
    D) mitosis.
    E) cytokinesis.
    Answer: C

36. The replication of DNA takes place during
    A) mitosis.
    B) meiosis I.
    C) meiosis II.
    D) interphase.
    E) cytokinesis.
    Answer: D

37. The assembly of microtubules that is responsible for the movement of chromosomes is the
    A) centromere.
    B) centrosome.
    C) chromatin.
    D) basal body.
    E) mitotic spindle.
    Answer: E

38. The splitting and separation of centromeres occurs during
    A) prophase.
    B) anaphase.
    C) metaphase.
    D) telophase.
    E) cytokinesis.
    Answer: B

39. The final stage of mitosis is
    A) prophase.
    B) interphase.
    C) anaphase.
    D) telophase.
    E) metaphase.
    Answer: D

40. The branch of medicine dealing with medical issues of aging is called:
    A) elderly physiology.
    B) gerontology.
    C) oncology.
    D) psychology.
    E) geriatrics.
    Answer: E

41. The study of tumors is called
    A) pathology.
    B) oncology.
    C) epidemiology.
    D) pharmacology.
    E) histology.
    Answer: B

42. The spread of cancerous cells to parts of the body is referred to as
    A) hyperplasia.
    B) metastasis.
    C) malignancy.
    D) mutation.
    E) hypertrophy.
    Answer: B

43. A cancer causing agent is a(n)
    A) oncogen.
    B) mutant.
    C) carcinogen.
    D) neoplasm.
    E) chemical.
    Answer: C

44. All of the following belong to abnormal cell division EXCEPT
    A) tumor.
    B) hyperplasia.
    C) neoplasm.
    D) cleavage furrow.
    E) benign tumor.
    Answer: D

45. Meiosis is the nuclear division used during the formation of
    A) sperm cells.
    B) skin cells.
    C) red blood cells.
    D) cancer cells.
    E) All of the above are correct.
    Answer: A


## True-False

*Write T if the statement is true and F if the statement is false.*

1. The term cytoplasm refers to all cellular contents between the plasma membrane and the nucleus.
   Answer: True

2. Peripheral proteins penetrate through the phospholipid bilayer of the plasma membrane.
   Answer: False

3. An electrochemical gradient exists between the outside and the inside of cells.
   Answer: True

4. Most proteins can easily pass through the plasma membrane.
   Answer: False

5. Fluid outside the cells of the body is called intracellular fluid.
   Answer: False

6. Kinetic energy is the energy of motion of molecules.
   Answer: True

7. Facilitated diffusion requires a membrane carrier and cellular energy.
   Answer: False

8. Osmotic pressure is the pressure needed to stop water movement across membranes.
   Answer: True

9. An eight percent NaCl solution is isotonic to most cells.
   Answer: False

10. The nucleus can communicate with the cytosol via nuclear pores.
    Answer: True

11. Ribosomes consist of three subunits of equal size.
    Answer: False

12. The Golgi complex produces lysosomes.
    Answer: True

13. Multiple projections on the surface of cells that move the mucus in the respiratory tract are called flagella.
    Answer: False

14. Human skin cells contain 23 chromosomes.
    Answer: False

15. The first stage of mitosis is called prophase.
    Answer: True

## Short Answer

*Write the word or phrase that best completes each statement or answers the question.*

1. The branch of science dealing with the study of cellular structure is _____ .
   Answer: cytology

2. The basic framework of the plasma membrane is the _____ _____.
   Answer: lipid bilayer

3. Proteins loosely attached to the interior surface of the cell membrane are called _____ proteins.
   Answer: peripheral

4. The fluid in the human body that is contained inside the cells is called _____ fluid.
   Answer: intracellular

5. When molecules move from an area of high concentration to an area of low concentration, they move with the ____.
   Answer: concentration gradient

6. When a solution has the same concentration of water molecules and solutes as a red blood cell, it is considered a(n) ____ solution.
   Answer: isotonic

7. The shrinkage that occurs when cells are placed into a hypertonic solution is called ____.
   Answer: crenation

8.  The movement of water through a selectively permeable membrane is _____.
    Answer: osmosis

9.  Molecules are transported from an area of lower to an area of higher concentration by ____.
    Answer: active transport

10. The projections of the plasma membrane formed by phagocytes around large solid particles outside the cell are ____.
    Answer: pseudopods

11. The thick, semifluid portion of the cytoplasm is the ____.
    Answer: cytosol

12. The nucleus is enclosed by the ____.
    Answer: nuclear envelope (membrane)

13. All proteins that are exported from the cell are released to the exterior of the cell by the process of ___.
    Answer: exocytosis

14. Cristae are a series of folds found in ____.
    Answer: mitochondria

15. Microfilaments, microtubules, and intermediate filaments are structures of the ____.
    Answer: cytoskeleton

16. Centrioles are cylindrical structures, which reside within the ____.
    Answer: centrosome

17. A group of nucleotides on a DNA molecule that codes for a particular protein is called a(n) ____.
    Answer: gene

18. A base triplet of AUG on mRNA would match the anticodon ____.
    Answer: UAC

19. Anticodons are located on ____.
    Answer: tRNA

20. The two major steps in protein synthesis are ____ and ____.
    Answer: transcription, translation

21. Meiosis is also called the ____ cell division
    Answer: reproductive

22. The division of the parent cell's cytoplasm is called ____.
    Answer: cytokinesis

23. A noncancerous growth is called a ____ tumor
    Answer: benign

24. A permanent structural change in a gene is a ____.
    Answer: mutation

25. Compounds that appear to prevent cellular damage associated with cancer, aging and heart disease are called _____.
    Answer: phytochemicals

## Matching

*Choose the item from Column 2 that best matches each item in Column 1.*

1. Column 1: A disease that effects the entire body
   Column 2: systemic disease

2. Column 1: The science that deals with why, when, and where diseases occur.
   Column 2: epidemiology

3. Column 1: The study that deals with the effects and uses of drugs
   Column 2: pharmacology

4. Column 1: The removal of tissue from the living body for diagnosis
   Column 2: biopsy

5. Column 1: The spread of cancer to other body tissues
   Column 2: metastasis

6. Column 1: Increase in the size of cells without cell division
   Column 2: hypertrophy

7. Column 1: A decrease in the size of cells
   Column 2: atrophy

## Essay

*Write your answer in the space provided or on a separate sheet of paper.*

1. Name five cell organelles and briefly describe the function of each.
   Answer: Golgi complex--sorts, modifies, and packages proteins and lipids: the Golgi complexes also
   form lysosomes.
   Ribosomes are the site of protein synthesis.
   Lysosomes contain digestive enzymes and digest cell substances and foreign microbes.
   Mitochondria are the site of ATP production.
   The nucleus is the control center of the cell.
   The rough endoplasmic reticulum is responsible for the production of proteins for export.
   The smooth endoplasmic reticulum is the site of lipid synthesis.

2. Draw and label a generalized animal cell. Include six structures.
   Answer: needs to be determined by the individual instructor

3. Define osmosis.
   Answer: Osmosis is the diffusion of water through a selectively permeable membrane.

4. Briefly describe the stages of mitosis
   Answer: 1. Prophase: the chromatin condenses and shortens into visible chromosomes, the
   nucleoli disappear, and the nuclear membrane breaks down. The mitotic spindle is
   formed during this stage of mitosis.
   2. Metaphase: the chromosomes line up on the metaphase plate (equatorial plate).
   3. Anaphase: the centromeres split, the sister chromatids separate, and each daughter
   chromosome moves toward the opposite poles of the cell.
   4. Telophase: chromosomal movement stops, microtubules disappear, a nuclear envelope
   reforms, and cytokinesis occurs.

## CHAPTER 4 Tissues

## Multiple-Choice

*Choose the one alternative that best completes the statement or answers the question.*

1. Which of the following tissues is avascular?
   A) connective tissue
   B) muscle tissue
   C) skeletal tissue
   D) epithelial tissue
   E) nervous tissue
   Answer: D

2. A scientist who examines tissue changes that might indicate disease is called
   A) histologist.
   B) radiologist.
   C) pathologist.
   D) cytologist.
   E) endocrinologist.
   Answer: C

3. The tissue lining body cavities is the
   A) epithelial tissue.
   B) connective tissue.
   C) skeletal tissue.
   D) muscle tissue.
   E) nervous tissue.
   Answer: A

4. The epithelial tissue which contains cells of different shapes and is capable of distention is
   A) simple columnar epithelium.
   B) pseudostratified epithelium.
   C) transitional epithelium.
   D) stratified cuboidal epithelium.
   E) squamous epithelium.
   Answer: C

5. Goblet cells are found in which of the following tissues?
   A) nervous tissue
   B) columnar epithelium
   C) cuboidal epithelium
   D) connective tissue
   E) none of the above
   Answer: B

6. The flat, single layered tissue which allows for diffusion to occur is
   A) pseudostratified epithelium.
   B) mesothelium.
   C) simple columnar epithelium.
   D) simple squamous epithelium.
   E) transitional tissue.
   Answer: D

7.  Which of the following tissues is often ciliated?
    A)  transitional epithelium
    B)  connective tissue
    C)  cartilage
    D)  columnar epithelium
    E)  squamous epithelium
    Answer: D

8.  The different types of epithelia are named according to
    A)  cell size.
    B)  location.
    C)  cell shape and arrangement of layers.
    D)  cell shape and location in the body.
    E)  location and function.
    Answer C

9.  Tissue that functions in support and protection of body organs, stores energy, and provides immunity is
    A)  epithelial tissue.
    B)  connective tissue.
    C)  nervous tissue.
    D)  muscle tissue.
    E)  osseous tissue.
    Answer: B

10. Which of the following epithelia function in absorption and secretion?
    A)  squamous epithelium
    B)  columnar epithelium
    C)  cuboidal epithelium
    D)  Both A and B
    E)  Both B and C
    Answer: E

11. The main function of stratified squamous epithelium is
    A)  secretion.
    B)  absorption.
    C)  diffusion.
    D)  protection.
    E)  distension.
    Answer: D

12. Epithelium that appears to have several layers of cells but does not, is classified as
    A)  simple epithelium.
    B)  stratified epithelium.
    C)  transitional epithelium.
    D)  pseudostratified epithelium.
    E)  columnar epithelium.
    Answer: D

13. Which of the following tissues contains a large amount of matrix?
    A)  connective tissue
    B)  glial tissue
    C)  epithelial tissue
    D)  muscle tissue
    E)  nervous tissue
    Answer: A

14. All of the following cells can be found in connective tissue EXCEPT
    A) fibroblasts.
    B) macrophages.
    C) plasma cell.
    D) mast cells.
    E) glial cells.
    Answer: E

15. Goblet cells are found in all of the following organ systems EXCEPT
    A) urinary system.
    B) endocrine system.
    C) digestive system.
    D) respiratory system.
    E) reproductive system.
    Answer: B

16. The fibers in the matrix of connective tissue are made of
    A) carbohydrates.
    B) proteins.
    C) nucleic acids.
    D) lipids.
    E) hyaluronic acid.
    Answer: B

17. Which of the following is responsible for waterproofing of the skin?
    A) hyaluronic acid
    B) chondroitin sulfate
    C) collagen
    D) keratin
    E) phospholipids
    Answer: D

18. A single layer of cuboidal cells lining a secretory duct would be classified as
    A) simple squamous epithelium.
    B) simple cuboidal epithelium.
    C) simple columnar epithelium.
    D) stratified cuboidal epithelium.
    E) stratified squamous epithelium.
    Answer: B

19. A basement membrane is always present
    A) between the epithelial cells.
    B) between the epithelium and the muscle tissue.
    C) between the epithelium and the connective tissue.
    D) on the apical surface of the epithelium.
    E) between the epithelium and the blood vessels.
    Answer: C

20. All of the following are secretory products of exocrine glands EXCEPT
    A) mucus.
    B) oil.
    C) digestive enzymes.
    D) hormones.
    E) saliva.
    Answer: D

21. The slippery, viscous substance that binds cells together and lubricates joints is
   A) hyaluronic acid.
   B) chondroitin sulfate.
   C) collagen.
   D) reticular connective tissue.
   E) elastin.
   Answer: A

22. Connective tissue fibers that can be stretched considerably without breaking are
   A) collagen fibers.
   B) reticular fibers.
   C) elastic fibers.
   D) glucoprotein fibers.
   E) chondroitin fibers.
   Answer: C

23. A type of connective tissue almost exclusively found in the embryo is
   A) loose connective tissue.
   B) mesenchyme.
   C) areolar connective tissue.
   D) vascular connective tissue.
   E) pericondrium.
   Answer: B

24. An embryonic connective tissue that supports the wall of the umbilical cord is
   A) mucous connective tissue.
   B) mesenchyme.
   C) areolar connective tissue.
   D) vascular connective tissue.
   E) elastic connective tissue.
   Answer: A

25. All of the following are classified as loose connective tissue EXCEPT
   A) areolar connective tissue.
   B) adipose tissue.
   C) reticular connective tissue.
   D) elastic connective tissue.
   E) Both C and D.
   Answer: E

26. To which of the following tissues does cartilage belong?
   A) epithelial tissue
   B) nervous tissue
   C) areolar tissue
   D) muscle tissue
   E) connective tissue
   Answer: E

27. The presence of chondrocytes and elastic fibers indicates
   A) dense connective tissue.
   B) hyaline cartilage.
   C) fibrocartilage.
   D) elastic cartilage.
   E) bone.
   Answer: D

28. The term "gristle" refers to
    A) hyaline cartilage.
    B) fibrocartilage.
    C) elastic cartilage.
    D) dense connective tissue.
    E) bone.
    Answer: A

29. What is the main function of fibrous connective tissue?
    A) absorption
    B) support
    C) contraction
    D) lining of body cavities
    E) secretion
    Answer: B

30. Blood belongs to which major tissue type?
    A) epithelial tissue
    B) connective tissue
    C) skeletal tissue
    D) muscle tissue
    E) nervous tissue
    Answer: B

31. Osseous tissue is classified as
    A) nervous tissue.
    B) cartilage.
    C) connective tissue.
    D) muscle tissue.
    E) epithelial tissue.
    Answer: B

32. All of the following are cells of connective tissue EXCEPT:
    A) fibroblasts.
    B) glial cells.
    C) mast cells.
    D) adipocytes.
    E) macrophages.
    Answer: B

33. Membranes that line body cavities that open directly to the exterior are
    A) serous membranes.
    B) synovial membranes.
    C) mucous membranes.
    D) parietal membranes.
    E) visceral membranes.
    Answer: C

34. Which of the following contain elastic connective tissue?
    A) heart valves
    B) the periosteum
    C) dermis
    D) epidermis
    E) wall of arteries
    Answer: E

35. The portion of a membrane that is closest to the organ is the
    A) mucous portion.
    B) parietal portion.
    C) synovial portion.
    D) visceral portion.
    E) pleural portion.
    Answer: D

36. The cells of mature cartilage are
    A) osteoblasts.
    B) osteocytes.
    C) chondroblasts.
    D) chondrocytes.
    E) fibroblasts.
    Answer: C

37. The basic unit of compact bone is a(n)
    A) osteon.
    B) trabeculae.
    C) osteocyte.
    D) osteoblast.
    E) canaliculus.
    Answer: A

38. The matrix of blood is
    A) serum.
    B) plasma.
    C) water.
    D) intracellular fluid.
    E) lymph.
    Answer: B

39. Intercalated discs are structures of
    A) smooth muscle.
    B) epithelial tissue.
    C) cardiac muscle.
    D) skeletal muscle.
    E) connective tissue.
    Answer: C

40. All of the following are structures of nervous tissue EXCEPT:
    A) neuroglia.
    B) axons.
    C) dendrite.
    D) cell body.
    E) periosteum.
    Answer: E

41. The membrane covering the lungs is the
    A) peritoneum.
    B) pleura.
    C) pericardium.
    D) mesothelium.
    E) endothelium.
    Answer: B

42. The membranes that line the cavities of some joints are
    A) synovial membranes.
    B) visceral membranes.
    C) mucous membranes.
    D) serous membranes.
    E) cutaneous membranes.
    Answer: A

43. The tissue that is highly specialized for contraction is
    A) nervous tissue.
    B) epithelial tissue.
    C) connective tissue.
    D) muscle tissue.
    E) osseous tissue.
    Answer: D

44. Neurolgia cells belong to which of the following tissues?
    A) epithelial tissue
    B) osseous tissue
    C) connective tissue
    D) muscle tissue
    E) nervous tissue
    Answer: E

45. Which of the following is an autoimmune disease?
    A) cystic fibrosis
    B) rheumatoid arthritis
    C) hypertension
    D) both A and B
    E) none of the above
    Answer: A


## True-False

*Write T if the statement is true and F if the statement is false.*

1.  A tissue is a group of similar cells that perform a specialized activity.
    Answer: True

2.  The lining of body cavities is one of the functions of connective tissue.
    Answer: False

3.  Epithelial cells are arranged in continuous sheets.
    Answer: True

4.  The exchange of materials between epithelium and connective tissue is an active transport mechanism.
    Answer: False

5.  Squamous epithelium has cube-shaped cells.
    Answer: False

6.  Epithelial tissue has a high mitotic capacity.
    Answer: True

7. Simple epithelium always consists of a layer of flat cells.
   Answer: False

8. Pseudostratified epithelium is never ciliated.
   Answer: False

9. Nonciliated columnar epithelium contains microvilli and goblet cells.
   Answer: True

10. Stratified columnar epithelium is a common type of tissue in the human body.
    Answer: False

11. Exocrine glands secrete their products into ducts.
    Answer: True

12. Cartilage is highly vascular.
    Answer: False

13. Connective tissue cells produce the ground substance of the connective tissue matrix.
    Answer: True

14. Mature connective tissue is not present in the newborn.
    Answer: False

15. Neuroglia protects and supports neurons.
    Answer: True

## Short Answer

*Write the word or phrase that best completes each statement or answers the question.*

1. The science that deals with the study of tissues is called ____.
   Answer: histology

2. Epithelium that consists of two or more layers of cells is ____ epithelium.
   Answer: stratified

3. The epithelium lining the urinary bladder is ____ epithelium.
   Answer: transitional

4. The pituitary gland is an example of a(n) _____ gland.
   Answer: endocrine

5. Exocrine glands are made out of _____ tissue.
   Answer: exocrine

6. Blood vessels are lined by _____.
   Answer: endothelium

7. A matrix and a cellular component are characteristics of ____ tissue.
   Answer: connective

8. The cells producing fibers in connective tissue are ____.
   Answer: fibroblasts

9. The most abundant connective tissue fibers are ____ fibers.
   Answer: collagen

10. Immature cells of cartilage are _____.
    Answer: chondroblasts

11. Blood is classified as ____ tissue.
    Answer: connective

12. Cells that are specialized for fat storage are ____.
    Answer: adipocytes

13. A connective tissue where bundles of collagen fibers have an orderly parallel arrangement is classified as ____ connective tissue.
    Answer: regular (dense regular)

14. The cells of mature cartilage are ____.
    Answer: chondrocytes

15. The type of cartilage that reduces friction at the joints is ____ cartilage.
    Answer: hyaline

16. Bone tissue is also called ____ tissue.
    Answer: osseous

17. The chemical that dilates small blood vessels and is released by mast cells is _____.
    Answer: histamine

18. The type of membrane in the body that does not contain epithelium is a(n) ____ membrane.
    Answer: synovial

19. The part of a serous membrane attached to the wall of the body cavity is the ____ portion.
    Answer: parietal

20. Synovial membranes are found in ____.
    Answer: joints

21. Cells that are capable of generating and conducting nerve impulses are ____.
    Answer: neurons (nerve cells)

22. The lining of the abdominal cavity is the ____.
    Answer: peritoneum

23. The covering of the heart is the ____.
    Answer: pericardium

24. The type of connective tissue that contains plasma is ____.
    Answer: blood

25. The procedure developed to grow skin in the laboratory is called ____ _____.
    Answer: tissue engineering

## Matching

*Choose the item from Column 2 that best matches each item in Column 1.*

1.  Column 1:  A tissue that can store fat
    Column 2:  Connective tissue

2.  Column 1:  A tissue that is avascular
    Column 2:  Epithelium

3.  Column 1:  A group of specialized cells which secrete their products into ducts
    Column 2:  Exocrine glands

4.  Column 1:  Fibers that consist of collagen and glycoproteins
    Column 2:  Reticular fibers

5.  Column 1:  An embryonic connective tissue
    Column 2:  Mesenchyme

6.  Column 1:  A type of loose connective tissue
    Column 2:  Adipose tissue

7.  Column 1:  Lines body cavities that do not open to the exterior
    Column 2:  Serous membranes


## Essay

*Write your answer in the space provided or on a separate sheet of paper.*

1. Name and briefly describe the function of the four major tissue types.
   Answer: 1. Epithelial tissue covers body surfaces and lines body cavities
   2. Connective tissue protects and supports the body and its organs, binds organs together, stores energy reserves, and provides immunity
   3. Muscular tissue is responsible for movement
   4. Nervous tissue initiates and transmits nerve impulses

2. Classify epithelia by shape and arrangement of cell layers.
   Answer:  simple squamous epithelium
   simple cuboidal epithelium
   simple columnar epithelium
   stratified squamous epithelium
   stratified cuboidal epithelium
   stratified columnar epithelium
   transitional epithelium
   pseudostratified columnar epithelium

3. Distinguish between exocrine and endocrine glands
   Answer:  Exocrine glands secrete their products into ducts, while endocrine glands secrete their products into the blood.

## CHAPTER 5   The Integumentary System

## Multiple-Choice

*Choose the one alternative that best completes the statement or answers the question.*

1.  A group of tissues that performs a specific function is a(n)
    A)  organ system
    B)  system
    C)  organ
    D)  specialized tissue
    E)  organism
    Answer: C

2.  The skin belongs to the
    A)  skeletal system
    B)  integumentary system
    C)  muscular system
    D)  sensory system
    E)  circulatory system
    Answer: B

3.  All of the following reflect homeostatic imbalances of the body EXCEPT
    A)  rashes
    B)  chicken pox
    C)  measles
    D)  cold sores
    E)  warts
    Answer: E

4.  The medical specialty that deals with the diagnosis and treatment of skin disorders is called
    A)  dermatology
    B)  pathology
    C)  pharmacology
    D)  cardiology
    E)  oncology
    Answer: A

5.  The outermost portion of the skin is the
    A)  hypodermis
    B)  subcutaneous layer
    C)  dermis
    D)  epidermis
    E)  basement membrane
    Answer: D

6.  All of the following are functions of the skin EXCEPT
    A)  protection
    B)  vitamin B synthesis
    C)  excretion
    D)  temperature regulation
    E)  sensation
    Answer: B

7. The protein that helps protect the skin and underlying cells is
   A) melanin
   B) melatonin
   C) keratin
   D) actin
   E) carotene
   Answer: C

8. The cells producing the pigment responsible for skin color are the
   A) keratinocytes
   B) melanocytes
   C) Langerhans cells
   D) Merkel cells
   E) Adipocytes
   Answer: B

9. Cells of the skin that are active in the immune process are
   A) keratinocytes
   B) melanocytes
   C) Merkel cells
   D) adipocytes
   E) Langerhans cells
   Answer: E

10. The layer of the epidermis which contains cells capable of continued cell division is the
    A) stratum spinosum
    B) stratum basale
    C) stratum corneum
    D) stratum granulosum
    E) stratum lucidum
    Answer: B

11. The layer of the epidermis typically found in the thick skin of the palms and soles is the
    A) stratum basale
    B) stratum spinosum
    C) stratum granulosum
    D) stratum lucidum
    E) stratum corneum
    Answer: D

12. The corpuscles of touch (Meissner's corpuscles) are found in the
    A) hypodermis
    B) dermis
    C) subcutaneous layer
    D) epidermis
    E) basement membrane
    Answer: B

13. All of the following are pigments involved in skin color EXCEPT
    A) keratin
    B) melanin
    C) carotene
    D) hemoglobin
    E) All of the above
    Answer: A

14. An inherited inability of an individual to produce melanin results in
    A) freckles
    B) melanoma
    C) erythema
    D) albinism
    E) carotenosis
    Answer: D

15. The redness of the skin that can occur during allergic reactions is
    A) albinism
    B) erythema
    C) due to carotene
    D) due to melanine
    E) caused by hemoglobin
    Answer: B

16. All of the following are accessory organs of the skin except
    A) hair
    B) nails
    C) oil glands
    D) Pacinian corpuscles
    E) sweat glands
    Answer: D

17. The primary function of hair is
    A) protection
    B) sensory perception
    C) to increase surface area
    D) temperature regulation
    E) movement
    Answer: A

18. The base of a hair follicle is enlarged into an onion-shaped structure called
    A) root
    B) shaft
    C) papilla
    D) bulb
    E) matrix
    Answer: D

19. Normal hair loss in the adult scalp is estimated at how many hairs per day?
    A) 200
    B) 100
    C) 50
    D) 1000
    E) 20
    Answer: B

20. The bundle of smooth muscles associated with hair follicles is called
    A) orbicularis muscle
    B) sebaceous bundle
    C) arrector pili
    D) goose bumps
    E) none of the above
    Answer: C

21. The glands usually associated with hair follicles are
    A) sudoriferous glands
    B) sweat glands
    C) ceruminous glands
    D) eccrine glands
    E) sebaceous glands
    Answer: E

22. Glands that are present in the external auditory meatus are
    A) ceruminous glands
    B) sudoriferous gland
    C) sebaceous glands
    D) eccrine
    E) apocrine glands
    Answer: A

23. Perspiration is the substance produced by
    A) ceruminous glands
    B) sudoriferous glands
    C) sebaceous glands
    D) oil glands
    E) holocrine glands
    Answer: B

24. The portion of a nail that extends past the end of the finger is the
    A) nail body
    B) nail root
    C) lunula
    D) free edge
    E) nail matrix
    Answer D

25. The cuticle of the nail consists of
    A) stratum corneum
    B) stratum lucidum
    C) nail matrix
    D) nail root
    E) basement membrane
    Answer: A

26. All of the following are true for the aging skin EXCEPT
    A) collagen fibers decrease in number
    B) elastic fibers loose some elasticity
    C) macrophages increase in number
    D) melanocytes increase in size
    E) fibroblasts decrease in number
    Answer: C

27. The layer of skin that contains dead cells filled with keratin is the
    A) stratum basale
    B) stratum lucidum
    C) stratum corneum
    D) stratum spinosum
    E) stratum granulosum
    Answer: C

28. Receptors in the skin that are sensitive to touch are
    A) dermal papillae
    B) chemoreceptors
    C) Pacinian corpuscles
    D) Meissner's corpuscles
    E) C and D are correct
    Answer: D

29. All of the following are necessary for the homeostasis of body temperature EXCEPT
    A) positive feedback
    B) the hypothalamus
    C) sweat glands
    D) blood vessels
    E) skeletal muscles
    Answer: A

30. A burn that involves the entire epidermis and some of the dermis is a
    A) first-degree burn
    B) second-degree burn
    C) third-degree burn
    D) fourth-degree burn
    E) blister
    Answer: A

31. An inflammation of sebaceous glands is called
    A) dermatitis
    B) decubitus
    C) ulcer
    D) acne
    E) burn
    Answer:  D

32. The most common form of skin cancer is
    A) basal cell carcinomas
    B) squamous cell carcinomas
    C) granular cell carcinomas
    D) malignant melanomas
    E) epidermal melanomas
    Answer: A

33. An inflammation of the skin is called
    A) hepatitis
    B) meningitis
    C) dermatitis
    D) abrasion
    E) hives
    Answer: C

34. Which of the following skin infections is caused by a fungus
    A) cold sores
    B) fever blisters
    C) hives
    D) Athlete's foot
    E) cysts
    Answer: D

35. Partial or complete loss of melanocytes from patches of skin causes a condition called
    A) albinism
    B) vitiligo
    C) keratosis
    D) laceration
    E) impetigo
    Answer: B

## True-False

*Write T if the statement is true and F if the statement is false.*

1. The hypodermis is the layer of skin below the epidermis.
   Answer: False

2. Langerhans' cells are important components in the immune system.
   Answer: True

3. The skin produces Vitamin D after exposure to UV light.
   Answer: True

4. The epidermis is the outermost layer of the skin.
   Answer: True

5. The stratum spinosum of the skin has a high mitotic activity.
   Answer: False

6. The dermis contains both elastic and collagen fibers.
   Answer: True

7. The lower layer of the epidermis contains adipose tissue.
   Answer: False

8. Carotene is a pigment of the skin.
   Answer: True

9. Carotene tends to form patches called freckles.
   Answer: False

10. Scalp hair grows continually throughout life.
    Answer: False

11. The matrix produces new hairs by cell division.
    Answer: True

12. Only some hair follicles go through a growth circle.
    Answer: False

13. Dark colored hair contains mostly true melanin.
    Answer: True

14. Sebum is secreted by sweat glands.
    Answer: False

15. Wrinkles are due to decrease of elasticity in elastic fibers of the skin.
    Answer: False

## Short Answer

*Write the word or phrase that best completes each statement or answers the question.*

1.  The pigment found in specialized cells of the stratum basale is ____.
    Answer: melanin

2.  Cold receptors are found in the ____.
    Answer: dermis

3.  Cancer of ____ is considered to be one of the most serious skin cancers.
    Answer: melanocytes

4.  The superficial portion of the hair is the ____.
    Answer: shaft

5.  The mammary glands are modified ____ glands.
    Answer: sudoriferous (sweat)

6.  Plates of tightly packed, hard, keratinized cells of the epidermis are referred to as ____.
    Answer: nails

7.  The portion of the nail that is visible is the ____.
    Answer: nail body

8.  The whitish portion of the nail is the ____.
    Answer: lunula

9.  The regulation of body temperature requires a ____ feedback system.
    Answer: negative

10. A burn that involves the surface of the epidermis only is a ____ -degree burn.
    Answer: first

11. Bedsores are ____ sores.
    Answer: pressure

12. Excessive exposure to the sun can result in ____.
    Answer: skin cancer

13. Cold sores are caused by a ____.
    Answer: virus

14. Medication that is applied to the surface is ____.
    Answer: topical

15. The layer below the dermis is called ____.
    Answer: hypodermis (subcutaneous layer)

16. Merkel cells are found in the ____.
    Answer: epidermis

17. In the upper layer of the dermis, finger-like extensions called ____ are present to increase the surface area.
    Answer: dermal papillae

18. Each hair follicle goes through a growth stage and a _____ stage.
    Answer: resting

19. The color of blackheads is due to _____ and _____.
    Answer: melanin, oxidized oil

20. Sweat glands associated with hair follicles are classified as _____ glands.
    Answer: apocrine

## Matching

*Choose the item from Column 2 that best matches each item in Column 1.*

1. Column 1: A portion of skin that has been scraped away.
   Column 2: Abrasion

2. Column 1: Caused by herpes simplex virus
   Column 2: Cold sore

3. Column 1: Condition of the skin marked by redness and often-itchy patches.
   Column 2: Hives

4. Column 1: Removal of tattoos by a high-speed brush.
   Column 2: Dermabrasion

5. Column 1: A superficial skin infection mostly in children caused by staphylococci or streptococci.
   Column 2: Impetigo

6. Column 1: A patch called a birthmark.
   Column 2: Nevus

7. Column 1: Formation of a hardened growth of tissue.
   Column 2: Keratosis

8. Column 1: Mass produced by uncontrolled growth of epithelial cells due to a virus.
   Column 2: Wart

9. Column 1: Wound or irregular tear of the skin.
   Column 2: Laceration

10. Column 1: A superficial fungal infection of the skin of the foot.
    Column 2: Athlete's feet

## Essay

*Write your answer in the space provided or on a separate sheet of paper.*

1. Name the functions of the skin.
   Answer: 1. Regulation of body temperature
   2. Protection
   3. Sensation
   4. Excretion
   5. Immunity
   6. Vitamin D synthesis

2. Name the layers of the skin starting at the deepest to the most superficial layer.
   Answer: 1. Stratum basale
   2. Stratum spinosum
   3. Stratum granulosum
   4. Stratum lucidum
   5. Stratum corneum

3. Name the glands associated with the skin and name the product each of these glands secretes.
   Answer: 1. Sebaceous glands (oil glands) secrete sebum, an oily substance
   2. Sudoriferous glands (sweat glands) secrete perspiration or sweat.
   3. Ceruminous glands secrete the combined product of sebaceous glands and ceruminous glands, which is called cerumen.

## CHAPTER 6 The Skeletal System

## Multiple-Choice

*Choose the one alternative that best completes the statement or answers the question.*

1.  All of the following are functions of the skeletal system EXCEPT
    A)  support.
    B)  blood cell production.
    C)  blood protein production.
    D)  mineral storage.
    E)  movement.
    Answer: C

2.  Which of the following is NOT considered to be a long bone?
    A)  bones of the ribcage
    B)  bones of the thighs
    C)  bones of the forearms
    D)  bones of the fingers
    E)  bones of the legs
    Answer: A

3.  Which of the following are classified as flat bones?
    A)  sesamoid bones
    B)  vertebrae
    C)  ankle bones
    D)  cranial bones
    E)  spongy bones
    Answer: D

4.  Yellow bone marrow
    A)  stores triglycerides.
    B)  consists of mainly adipose cells.
    C)  produces red blood cells.
    D)  A and B are correct.
    E)  A and C are correct.
    Answer: D

5.  Small bones located within special joints are called
    A)  flat bones.
    B)  sutural bones.
    C)  irregular bones.
    D)  long bones.
    E)  short bones.
    Answer: B

6.  The shaft of a long bone is called
    A)  epiphysis.
    B)  diaphysis.
    C)  metaphysis.
    D)  periosteum.
    E)  endosteum.
    Answer: B

7. The site where bone growth occurs is the
    A) epiphysial plate.
    B) epiphysis.
    C) diaphysis.
    D) articular cartilage.
    E) periosteum.
    Answer: A

8. The endosteum
    A) contains red bone marrow.
    B) is the unit of compact bone.
    C) lines the medullary cavity.
    D) surrounds the bone surface.
    E) is the end of long bones.
    Answer: C

9. Which of the following cells in osseous tissue has mitotic potential?
    A) osteoblast
    B) osteoclast
    C) osteocyte
    D) bone cell
    E) osteoprogenitor cell
    Answer: E

10. Osteogenic cells
    A) are mature bone cells.
    B) develop into osteoblasts.
    C) break down bone.
    D) deposit calcium into the bone.
    E) surround themselves with matrix.
    Answer: B

11. Osteoclasts
    A) deposit calcium into bone.
    B) destroy bone matrix.
    C) are a type of white blood cell.
    D) A and C are correct.
    E) None of the above are correct.
    Answer: D

12. Nutrient arteries and nerves from the periosteum penetrate compact bone through the
    A) Haversian canal.
    B) central canal.
    C) Volkman's canal.
    D) concentric lamellae.
    E) canaliculi.
    Answer: C

13. Fat is stored in
    A) compact bone.
    B) spongy bone.
    C) red bone marrow.
    D) yellow bone marrow.
    E) articular cartilage.
    Answer: D

14. The secondary ossification center develops
    A) during intramembranous ossification.
    B) after development of the trabeculae.
    C) when blood vessels enter the epiphyses.
    D) at the appearance of the epiphyseal line.
    E) simultaneously with the primary ossification center.
    Answer: C

15. The cells responsible for the re-absorption of bone are
    A) osteoprogenitor cells.
    B) osteoclasts.
    C) osteocytes.
    D) osteoblasts.
    E) bone forming cells.
    Answer: B

16. The hormone(s) responsible for the homeostasis of blood calcium levels is (are)
    A) calcitonin.
    B) thyroid hormones.
    C) insulin and glucagon.
    D) parathyroid hormone and calcitonin.
    E) parathyroid hormone only.
    Answer: D

17. The rate of bone loss during aging can be slowed down by
    A) regular walking.
    B) rest and diet.
    C) diet only.
    D) vitamin injections.
    E) steroid treatment.
    Answer: A

18. All of the following hormones play a role in bone growth and maintenance EXCEPT:
    A) human growth hormone.
    B) parathyroid hormone.
    C) insulin.
    D) thyroid hormones.
    E) androgens.
    Answer: B

19. Which of the following substances does NOT have a known effect on bone metabolism?
    A) vitamin D
    B) vitamin E
    C) vitamin A
    D) calcium
    E) phosphorus
    Answer: B

20. An opening in bone through which blood vessels and nerves pass is a
    A) depression.
    B) tuberosity.
    C) sinus.
    D) meatus.
    E) foramen.
    Answer: E

21. A depression in or on a bone is a
    A) fossa.
    B) tuberosity.
    C) foramen.
    D) condyle.
    E) meatus.
    Answer: A

22. The adult human skeleton consists of
    A) 230 bones.
    B) 200 bones.
    C) 206 bones.
    D) 198 bones.
    E) 302 bones.
    Answer: C

23. All of the following are facial bones EXCEPT
    A) nasal bone.
    B) maxilla.
    C) mandible.
    D) ethmoid bone.
    E) vomer.
    Answer: D

24. Which of the following is a cranial bone?
    A) vomer bone
    B) lacrimal bone
    C) zygomatic bone
    D) sphenoid bone
    E) palatine bone
    Answer: D

25. The suture between the parietal bones and the temporal bones is the
    A) sagittal suture.
    B) parasaggittal suture.
    C) lamboid suture.
    D) coronal suture.
    E) squamous suture.
    Answer: E

26. The temporal and zygomatic bones join to form the
    A) zygomatic arch.
    B) mandibular fossa.
    C) foramen magnum.
    D) mastoid process.
    E) styloid process.
    Answer: A

27. The medulla oblongata passes through the
    A) mental foramen.
    B) foramen magnum.
    C) occipital condyles.
    D) foramen ovale.
    E) carotid foramen.
    Answer: B

28. Which of the following forms the roof of the nasal cavity?
    A) superior nasal concha
    B) inferior nasal concha
    C) cribriform plate
    D) perpendicular plate
    E) nasal septum
    Answer: C

29. The upper jawbone is formed by the
    A) alveoli.
    B) maxillae.
    C) mandible.
    D) zygomatic bones.
    E) lacrimal bones.
    Answer: B

30. Which of the following is the only movable bone of the skull?
    A) maxilla
    B) zygomatic bone
    C) lacrimal bone
    D) mandible
    E) hyoid bone
    Answer: D

31. Technically, the cheekbones are called
    A) zygomatic bones.
    B) lacrimal bones.
    C) maxillae.
    D) vomer bones.
    E) frontal bones.
    Answer: A

32. The bone that does not articulate with any other bone is the
    A) vomer bone.
    B) zygomatic bone.
    C) lacrimal bone.
    D) ethmoid bone.
    E) hyoid bone.
    Answer: E

33. The adult vertebral column contains
    A) 32 vertebra.
    B) 30 vertebra.
    C) 26 vertebra.
    D) 24 vertebra.
    E) 20 vertebra.
    Answer: C

34. A spinal nerve leaves the vertebral column at the
    A) interventricular foramen.
    B) intervertebral foramen.
    C) vertebral arch.
    D) vertebral foramen.
    E) foramen magnum.
    Answer: B

35. Which of the following terms do NOT belong to a vertebrum?
    A) transverse process
    B) lateral process
    C) spinous process
    D) superior articular process
    E) pedicles
    Answer: B

36. The largest and strongest of the vertebrae in the spinal cord are the
    A) cervical vertebrae.
    B) thoracic vertebrae.
    C) lumbar vertebrae.
    D) sacral vertebrae.
    E) coccygeal vertebrae.
    Answer: C

37. Which of the following vertebra articulate with the ribs?
    A) sacral vertebra
    B) lumbar vertebra
    C) cervical vertebra
    D) thoracic vertebra
    E) coccygeal vertebra
    Answer: D

38. All of the following are bones of the thorax EXCEPT
    A) clavicle.
    B) sternum.
    C) true ribs.
    D) thoracic vertebrae.
    E) false ribs.
    Answer: A

39. The pectoral girdle consists of the
    A) humerus and radius.
    B) ulna and radius.
    C) hipbones.
    D) sternum and ribs.
    E) clavicle and scapula.
    Answer: E

40. The bone of the upper arm is the
    A) radius.
    B) ulna.
    C) humerus.
    D) femur.
    E) tibia.
    Answer: C

41. All of the following are carpals EXCEPT
    A) cuboid.
    B) lunate.
    C) hamate.
    D) capitate.
    E) scaphoid.
    Answer: A

42. The anatomical term for shinbone is
    A) fibula.
    B) tibia.
    C) humerus.
    D) femur.
    E) ulna.
    Answer: B

43. The heaviest bone in the body is the
    A) humerus.
    B) fibula.
    C) tibia.
    D) hipbone.
    E) femur.
    Answer: E

44. Which of the following is NOT a bone of the foot?
    A) navicular.
    B) calcaneus.
    C) cuboid.
    D) talus.
    E) pisiform.
    Answer: E

45. Which of the following is a disorder in which bone calcification fails?
    A) osteoporosis
    B) herniated disc
    C) osteopenia
    D) osteomalacia
    E) scoliosis
    Answer: D

## True-False

*Write T if the statement is true and F if the statement is false.*

1. The brain is protected by cranial bones.
   Answer: True

2. Vertebrae are also called Wormian bones.
   Answer: False

3. The medullary cavity contains red bone marrow.
   Answer: False

4. The endosteum is the lining of the medullary cavity.
   Answer: True

5. Osteoblasts are mature bone cells.
   Answer: False

6. Canaliculi are characteristic of spongy bone.
   Answer: False

7. Bone formation is called ossification.
   Answer: True

8. The replacement of cartilage by bone is called intramembranous ossification.
   Answer: False

9. The epiphyseal plate allows the diaphysis of the bone to increase in diameter.
   Answer: False

10. Bone continuously replaces itself throughout adult life.
    Answer: True

11. The decrease of estrogen levels in females after the age of 40 increases the loss of calcium from bones.
    Answer: True

12. The sphenoid bone is one of the facial bones.
    Answer: False

13. The vomer is a triangular bone, which forms part of the nasal septum.
    Answer: True

14. The openings between the vertebrae form the vertebral canal.
    Answer: False

15. The shoulder blades are also called clavicles.
    Answer: False


## Short Answer

*Write the word or phrase that best completes each statement or answers the question.*

1. Hemopoiesis occurs in _____ _____ _____.
   Answer: red bone marrow

2. The distal and proximal ends of long bones are called _____.
   Answer: epiphyses

3. The tough, white, fibrous membrane around the surface of a bone is called ____.
   Answer: periosteum

4. Trabeculae are found in ____ bone.
   Answer: spongy

5. The formation of bone within fibrous connective tissue membranes is referred to as _____.
   Answer: intramembranous ossification

6. The ongoing replacement of old bone tissue by new bone tissue is called bone _____.
   Answer: remodeling

7. The ribs belong to the ____ skeleton.
   Answer: axial

8. The minerals responsible for the hardening of bone matrix are _____ and _____.
   Answer: calcium and phosphorus

9. Elderly people can strengthen their bones by _____-_____ exercise.
   Answer: weight-bearing

10. An immovable joint between the bones of the skull is called a ____.
    Answer: suture

11. The suture between the frontal bone and the two parietal bones is the _____ suture.
    Answer: coronal

12. The bony sockets into which teeth are set are called ____.
    Answer: alveoli

13. The bone marking used as a landmark by dentists to inject anesthetics is the ____.
    Answer: mental foramen

14. The smallest facial bones are the ____ bones.
    Answer: lacrimal

15. The correct anatomical term for "soft spots" on infant skulls is ____.
    Answer: fontanels

16. The structures found between adjacent vertebra are ____.
    Answer: intervertebral discs

17. The first cervical vertebra is called ____.
    Answer: atlas

18. The medial bone of the forearm is the ____.
    Answer: ulna

19. The bones of the fingers are called the ____.
    Answer: phalanges

20. The hipbones are united to each other by a joint called ____.
    Answer: pubic symphysis

21. The correct anatomical term for kneecap is ____.
    Answer: patella

22. The collective term for the seven bones of the ankle is ____.
    Answer: tarsus

23. The disorder characterized by decrease in bone mass is called ____.
    Answer: osteoporosis

24. A lateral bending of the vertebral column is called ____.
    Answer: scoliosis

25. A deformity of the great toes is called ____.
    Answer: bunion

## Matching

*Choose the item from Column 2 that best matches each item in Column 1.*

1.  Column 1: A Vitamin D deficiency in children in which the body does not absorb calcium and
               phosphorus.
    Column 2: Rickets

2. Column 1: An irregular thickening and softening of the bones.
Column 2: Paget's disease

3. Column 1: A lateral bending of the vertebral column, usually the thoracic spine.
Column 2: Scoliosis

4. Column 1: A deformity of the great toe.
Column 2: Bunion

5. Column 1: A disorder characterized by decreased bone mass.
Column 2: Osteoporosis

6. Column 1: Mastoid process
Column 2: Temporal bones

7. Column 1: Carotid foramen
Column 2: Temporal bones

8. Column 1: Sella turcica
Column 2: Sphenoid bone

9. Column 1: Optic foramen
Column 2: Sphenoid bone

10. Column 1: Condylar process
Column 2: Mandible

11. Column 1: Spinous process
Column 2: Vertebrum

12. Column 1: Acromion
Column 2: Scapula

13. Column 1: Coronoid fossa
Column 2: Humerus

14. Column 1: Acetabulum
Column 2: Hipbone

15. Column 1: Foramen magnum
Column 2: Occipital bone

## Essay

*Write your answer in the space provided or on a separate sheet of paper.*

1. Explain the difference between true ribs, false ribs, and floating ribs.
Answer:  True ribs have a direct anterior attachment to the sternum via the costal cartilage.
The costal cartilage of false ribs does not directly attach to the sternum, and
floating ribs have no anterior attachment to the sternum at all.

2. Name the portions of the sternum.
Answer:  Manubrium, body, and xiphoid process.

3. List the cranial bones.
Answer:  frontal bone, parietal bones (2), temporal bones (2), occipital bone, sphenoid bone, ethmoid
bone

4.  Name the eight carpal bones.
    Answer:  scaphoid (navicular), lunate, triquetrum, pisiform, trapezium, trapezoid, capitate, hamate

## CHAPTER 7 Joints

## Multiple-Choice

*Choose the one alternative that best completes the statement or answers the question.*

1. The scientific study of joints is called
   A) anthropology.
   B) archeology.
   C) arthrology.
   D) orthoscopy.
   E) rheumatology.
   Answer: C

2. All of the following are structural classifications of joints EXCEPT
   A) synarthrosis.
   B) fibrous.
   C) cartilaginous.
   D) synovial.
   E) All of the above are correct.
   Answer: A

3. Slightly movable joints are called
   A) synovial joints.
   B) amphiarthorses.
   C) synarthrosis.
   D) sutures.
   E) diarthroses.
   Answer: B

4. Which of the following joints is classified as a synarthrotic joint?
   A) sydesmoses
   B) diarthrosis
   C) symphysis
   D) gomphosis
   E) amphiarthrosis
   Answer: D

5. The articulation between the root of the teeth and the alveoli is an example of a
   A) synchondrosis.
   B) diarthrosis.
   C) amphiarthrosis.
   D) suture.
   E) gomphosis.
   Answer: E

6. A joint in which the connecting material is a broad flat, disc of fibrocartilage is a
   A) symphysis.
   B) syndemosis.
   C) synchondrosis.
   D) gomphosis.
   E) suture.
   Answer: A

7. Which of the following is a cartilaginous joint?
   A) diarthrosis
   B) symphyses
   C) syndesmosis
   D) suture
   E) gomphosis
   Answer: B

8. Freely moveable joints are called
   A) synarthroses.
   B) diarthroses.
   C) symphysis.
   D) amphiarthroses.
   E) gomphosis.
   Answer: B

9. A cavity between joints is characteristic for
   A) synovial joints.
   B) symphysis.
   C) fibrous joints.
   D) amphiathrosis.
   E) None of the above.
   Answer: A

10. The inner layer of the articular capsule is a
    A) fibrous membrane.
    B) ligament.
    C) articular membrane.
    D) synovial membrane.
    E) meniscus.
    Answer: D

11. All of the following are functions of the synovial fluid EXCEPT
    A) lubrication.
    B) friction reduction.
    C) protein production.
    D) removal of wastes.
    E) supply of nutrients.
    Answer: C

12. A tearing of the menisci of the knee is commonly called a torn
    A) ligament.
    B) bursa.
    C) cartilage.
    D) tendon.
    E) muscle.
    Answer: C

13. Sac-like structures that reduce friction between moving parts are called
    A) tendons.
    B) ligaments.
    C) menisci.
    D) accessory ligaments.
    E) bursae.
    Answer: E

14. A gliding movement is a movement of a(n)
    A) fibrous joint.
    B) cartilaginous joint.
    C) synovial joint.
    D) All of the above.
    E) None of the above.
    Answer: C

15. The joints between carpals are examples of a
    A) ball-and -socket joint.
    B) hinge joint.
    C) pivot joint.
    D) gliding joint.
    E) saddle joint.
    Answer: D

16. Which of the following is/are synovial joints?
    A) planar joints
    B) hinge joints
    C) condyloid joints
    D) none of the above
    E) all of the above
    Answer: E

17. A movement which decreases the angle between articulating bones is called a(n)
    A) flexion.
    B) extension.
    C) circumduction.
    D) adduction.
    E) abduction.
    Answer: A

18. The joint between the trochlea of the humerus and the trochlear notch of the ulna at the elbow is an example of a
    A) saddle joint.
    B) pivot joint.
    C) hinge joint.
    D) condyloid joint.
    E) gliding joint.
    Answer: C

19. Extension is a movement at a
    A) gliding joint.
    B) hinge joint.
    C) pivot joint.
    D) ball-and-socket joint.
    E) saddle joint.
    Answer: B

20. The primary movement of a pivot joint is
    A) rotation.
    B) abduction.
    C) adduction.
    D) circumduction.
    E) hyperextension.
    Answer: A

21. The joint between the radius and the carpals is an example of a
    A) saddle joint.
    B) pivot joint.
    C) ball-and socket-joint.
    D) hinge joint.
    E) condyloid joint.
    Answer: E

22. Side-to-side and back-and -forth movement is typical for a
    F) saddle joint.
    G) pivot joint.
    H) hinge joint.
    I) condyloid joint.
    J) gliding joint.
    Answer: A

23. Up-and-down and side-to-side movements are performed by
    A) saddle joint.
    B) pivot joint.
    C) hinge joint.
    D) condyloid joint.
    E) gliding joint.
    Answer: D

24. The shoulder joint is an example of a
    A) saddle joint.
    B) pivot joint.
    C) ball-and-socket joint.
    D) hinge joint.
    E) gliding joint.
    Answer: C

25. The lateral movement of the arms away from the body is called
    A) elevation.
    B) abduction.
    C) adduction.
    D) flexion.
    E) pronation.
    Answer: B

26. Which of the following is NOT considered a special movement only occurring at particular joints?
    A) elevation
    B) protraction
    C) circumduction
    D) supination
    E) dorsiflexion
    Answer: C

27. Hyperextension represents
    A) the movement of a bone away from the midline.
    B) the movement of the distal end of a part of the body in a circle.
    C) the movement of a part of the body forward.
    D) an extension beyond the anatomical position.
    E) movement of a protracted part of the body back to the anatomical position.
    Answer: D

28. The shrugging of the shoulders to elevate the scapula is an example of
   A) abduction.
   B) depression.
   C) inversion.
   D) circumduction.
   E) elevation.
   Answer: E

29. The ligament that extends from the femur to the tibia is the
   A) patellar ligament.
   B) arcuate popliteal ligament.
   C) oblique popliteal ligament.
   D) medial meniscus.
   E) tibial ligament.
   Answer: C

30. An autoimmune disease that attacks the joints is
   A) rheumatoid arthritis.
   B) Lyme disease.
   C) osteoarthritis.
   D) arthralgia.
   E) multiple sclerosis.
   Answer: A

31. The field of medicine devoted to the disease of the joints is
   A) oncology.
   B) arthroscopy.
   C) dermatology.
   D) radiology.
   E) rheumatology.
   Answer: E

32. The arcuate popliteal ligament is found in the
   A) sole of the foot.
   B) knee.
   C) elbow.
   D) hip.
   E) shoulder.
   Answer: B

33. An inflammation of the joints is called
   A) bursitis.
   B) tendonitis.
   C) arthritis.
   D) perichondritis.
   E) chondritis.
   Answer: C

34. Lyme disease is caused by
   A) fever.
   B) a virus.
   C) ticks.
   D) a bacterium.
   E) All of the above.
   Answer: D

35. What type of cartilage is articular cartilage?
   A) fibrous
   B) reticular
   C) hyaline
   D) elastic
   E) collagen
   Answer: C

36. The outer layer of the articular capsule consists of
   A) dense irregular connective tissue.
   B) dense regular connective tissue.
   C) cartilage.
   D) synovial membranes.
   E) loose connective tissue.
   Answer: A

37. A decrease of the angles between bones is achieved by
   A) flexion.
   B) extension.
   C) adduction.
   D) abduction.
   E) pronation.
   Answer: A

38. A forward movement of the mandible would be called
   A) inversion.
   B) protraction.
   C) flexion.
   D) retraction.
   E) supination.
   Answer: B

39. Spur (new Bone) formation is characteristic of which disorder
   A) Lyme disease.
   B) rheumatoid arthritis.
   C) gouty arthritis.
   D) osteoarthritis.
   E) chondritis.
   Answer: D

40. Bending the foot upward involves
   A) elevation.
   B) planar flexion.
   C) dorsiflexion.
   D) inversion.
   E) supination.
   Answer: C

41. All of the following are part of the knee joint EXCEPT
   A) medial meniscus.
   B) patellar ligament.
   C) articular capsule.
   D) bursae.
   E) transverse humeral ligament.
   Answer: E

42. A joint that is eventually replaced by bone is a
    A) synchondrosis.
    B) sydesmoisis.
    C) symphysis.
    D) diarthrosis.
    E) suture.
    Answer: A

43. The distal articulation of the tibia and fibula is a
    A) diarthrosis.
    B) sydesmoisis.
    C) synchondrosis.
    D) symphysis.
    E) synarthrosis.
    Answer: B

44. The ligament from the femur to the tibia that strengthens the posterior surface of the joint is the
    A) oblique poplietal ligament.
    B) patellar ligament.
    C) tibial collateral ligament.
    D) posterior cruciate ligament.
    E) fibular collateral ligament.
    Answer: A

45. The displacement of a bone from a joint with tearing of ligaments, tendons, and articular capsules is called a(n)
    A) synovitis.
    B) chondritis.
    C) arthralgia.
    D) dislocation.
    E) sprain.
    Answer: D

## True-False

*Write T if the statement is true and F if the statement is false.*

1. The contact between teeth and bone is an articulation.
   Answer: True

2. A suture is a fibrous joint.
   Answer: True

3. Synarthroses are slightly moveable joints.
   Answer: False

4. Pads of fibrocartilage inside the knee joint are called articular discs.
   Answer: True

5. The epiphyseal plate is an example of a gomphosis.
   Answer: False

6. Ligaments are present in all types of articulations.
   Answer: True

7. Abduction refers to movement away from the midline of the body.
   Answer: True

8. The posterior cruciate ligament extends posteriorly from the tibia to the femur.
   Answer: False

9. Arthritis is a form of rheumatism.
   Answer: True

10. A strain is more serious than a sprain.
    Answer: False

11. An incomplete dislocation is called sublaxation.
    Answer: True

12. Arthralgia is the displacement of a bone from a joint.
    Answer: False

13. The anterior cruciate ligament belongs to the shoulder joint.
    Answer: False

14. Pronation is a movement of the forearm in which the palm is turned backward or downward.
    Answer: True

15. Syndesmoses are slightly moveable joints.
    Answer: True

## Short Answer

*Write the word or phrase that best completes each statement or answers the question.*

1. The field of medicine dealing with joint diseases is _____.
   Answer: rheumatology

2. Another name for *joint* is _____.
   Answer: articulation

3. The study of the movement of the human body is called _____.
   Answer: kinesiology

4. The space between articulating bones is called ____.
   Answer: synovial cavity

5. The displacement of the articulating bones from their normal position is called _____.
   Answer: dislocation

6. The three types of fibrous joints are sutures, syndesmoses, and _____.
   Answer: gomphoses

7. A tearing of menisci in the knee is commonly called _____ _____.
   Answer: torn cartilage

8. A freely movable joint is classified as a(n) ____.
   Answer: diarthrosis

9. The fibrous joints between the bones of the skull are called ____.
   Answer: sutures

10. The epiphyseal plate is a type of immovable joint (synarthrosis) called ____.
    Answer: synchondrosis

11. The distal articulation of the tibia and fibula is a slightly moveable joint (amphiarthrosis) of the type ____.
    Answer: syndesmosis

12. The capsule that surrounds a synovial cavity is called ____.
    Answer: articular capsule

13. The synovial membrane secretes ____.
    Answer: synovial fluid

14. An inflammation of the bursa is called ____.
    Answer: bursitis

15. The type of freely movable joints in the ankle are ____ joints.
    Answer: hinge

16. A movement toward the midline of the body is called a(n) ____.
    Answer: adduction

17. An upward movement of a part of the body is a(n) ____.
    Answer: elevation

18. The lateral meniscus belongs to the ____ joint.
    Answer: knee

19. The displacement of a bone from a joint with the tearing of ligaments, tendons, and articular capsule is called a ____.
    Answer: dislocation or luxation

20. The inflammation of cartilage is ____.
    Answer: chondritis

21. The point of contact between two or more bones is called a(n) ____.
    Answer: articulation

22. Structurally, joints are classified as synovial, fibrous, and _____.
    Answer: cartilaginous

23. Functionally, joints are classified as synarthroses, diarthroses, and ____.
    Answer: amphiarthroses

24. The specific type of synovial joints between the radius and the ulna is a ____ joint.
    Answer: pivot

25. The specific classification given the synovial joint of the shoulder, is an example of a(n) _____ joint.
    Answer: ball-and-socket

## Matching

*Choose the item in Column 2 that best matches each item in Column 1.*

1. Column 1: An immovable joint.
   Column 2: synarthrosis

2. Column 1: A freely movable joint.
   Column 2: diarthrosis

3. Column 1: A fibrous joint found between the bones of the skull.
   Column 2: suture

4. Column 1: A joint that permits movements in three planes.
   Column 2: ball-and-socket joint

5. Column 1: A joint that is capable of hyperextension.
   Column 2: hinge joint

6. Column 1: A joint between the sternum and the clavicle.
   Column 2: gliding joint

7. Column 1: A joint that provides sliding or twisting movement.
   Column 2: gliding joint

## Essay

*Write your answer in the space provided or on a separate sheet of paper.*

1. Classify the joints according to function and indicate the degree of movement each permits.
   Answer: Synarthroses are immovable joints, amphiarthroses are slightly movable joints, and Diarthroses are freely moveable joints.

2. Name the different types of diarthroses.
   Answer: gliding joints, hinge joints, pivot joints, condyloid joints, saddle joints, ball-and-socket joints

3. Name and describe the six special movements that relax the foot and hand.
   Answer: Inversion is the movement of the sole inward, eversion is the movement of the sole outward, dorsiflexion involves bending the foot downward, plantar flexion involves pointing the toes, and supination and pronation are movements of the forearm.

## CHAPTER 8   The Muscular System

## Multiple-Choice

*Choose the one alternative that best completes the statement or answers the question.*

1. All of the following are characteristics of muscle tissue EXCEPT
   A) transmissibility.
   B) contractility.
   C) extensibility.
   D) excitability.
   E) elasticity.
   Answer: A

2. Which of the following is NOT a function of muscle tissue?
   A) Heat production
   B) Regulation of organ volume
   C) Vitamin D production
   D) Body movements
   E) Body position
   Answer: C

3. The structure made of dense irregular connective tissue that holds muscles together and separates them into functional groups is
   A) epimysium.
   B) endomysium.
   C) perimysium.
   D) superficial fascia.
   E) deep fascia.
   Answer: E

4. Bundles of muscle fibers are covered by
   A) superficial fascia.
   B) deep fascia.
   C) perimysium.
   D) epimysium.
   E) endomysium.
   Answer: C

5. The portion of the sarcomere mostly composed of thick filaments is the
   A) H zone.
   B) Z discs.
   C) I band.
   D) A band.
   E) sarcomere.
   Answer: D

6. The sarcomere is the area between two
   A) Z discs.
   B) I bands.
   C) A bands.
   D) H zones.
   E) B lines.
   Answer: A

7. The pigment in the muscle fibers that stores oxygen is
   A) hemoglobin.
   B) myoglobin.
   C) melanin.
   D) melanin.
   E) glycogen.
   Answer: B

8. The neurotransmitter at the neuromuscular junction is
   A) norepinephrine.
   B) acetylcholine.
   C) adrenaline.
   D) calcium.
   E) glycine.
   Answer: B

9. Tunnel-like extensions of the sarcolemma into the muscle fiber are
   A) sarcomeres.
   B) myofibrils.
   C) transverse tubules.
   D) tropomyosin.
   E) myosin cross bridges.
   Answer: C

10. The structure storing calcium in the skeletal muscle fibers at rest is the
    A) endoplasmic reticulum.
    B) nucleus.
    C) Golgi apparatus.
    D) sarcoplasmic reticulum.
    E) T –Tubule.
    Answer: D

11. All of the following are proteins of myofilaments EXCEPT
    A) actin.
    B) myosin.
    C) troponin.
    D) tropomyosin.
    E) elastin.
    Answer: E

12. The space between an axon terminal and the sarcolemma is the
    A) neuromuscular junction.
    B) synaptic cleft.
    C) synaptic vesicle.
    D) motor end plate.
    E) motor unit.
    Answer: B

13. Which of the following substances can block the release of acetylcholine at the neuromuscular junction?
    A) acetylcholinesterase
    B) curare
    C) Botulinum toxin
    D) Anticholinesterases
    E) Cocaine
    Answer: C

14. Which of the following substances are absolutely necessary for muscle contraction?
    A) potassium
    B) ATP and calcium
    C) ATP and sodium
    D) ADP and phosphate
    E) potassium and ATP
    Answer: B

15. Acetylcholine in the synaptic cleft is broken down by
    A) anticholinesterase.
    B) pseudocholinesterese.
    C) acetylcholinesterase.
    D) cholinesterase.
    E) None of the above
    Answer: C

16. Muscle tone is due to
    A) the contraction of most muscle fibers in a whole muscle.
    B) a small number of motor units activated.
    C) sustained muscle contraction.
    D) None of the above.
    E) All of the above.
    Answer: B

17. Muscle fatigue can be caused by
    A) increased ADP.
    B) insufficient oxygen.
    C) depletion of creatine phosphate.
    D) A and B.
    E) All of the above.
    Answer: E

18. Most of the lactic acid produced by anaerobic respiration in muscle fibers is reconverted to glucose in the
    A) muscle fibers.
    B) liver.
    C) pancreas.
    D) kidney.
    E) blood.
    Answer: B

19. A sustained contraction of a muscle is called
    A) tetanus.
    B) summation.
    C) latent period
    D) twitch.
    E) refractory period.
    Answer: A

20. An increase in the diameter of muscle fibers is called
    A) muscular atrophy.
    B) muscular hypertrophy.
    C) hypertonia.
    D) hypotonia.
    E) Both A and D are correct.
    Answer: B

21. Which of the following is/are striated and involuntary?
    A) skeletal muscle
    B) cardiac muscle
    C) smooth muscle
    D) both a and b are correct
    E) both a and c are correct
    Answer: B

22. Intercalated discs are characteristic structures in
    A) cardiac muscle.
    B) smooth muscle.
    C) skeletal muscle.
    D) any striated muscle.
    E) striated and smooth muscle.
    Answer A

23. Intermediate filaments are structures of
    A) cardiac muscle.
    B) smooth muscle.
    C) skeletal muscle.
    D) striated muscle.
    E) skeletal and smooth muscle.
    Answer: B

24. Inflammation of muscle fibers is called
    A) myalgia.
    B) myoma.
    C) myositis.
    D) paralysis.
    E) myospasm.
    Answer: C

25. The ability of muscle fibers to shorten is referred to as
    A) excitability.
    B) contractility.
    C) extensibility.
    D) elasticity.
    E) recoiling.
    Answer: B

26. The immediate direct source of energy for muscle contraction is
    A) creatine phosphate.
    B) myoglobin.
    C) ATP.
    D) ADP.
    E) glycogen.
    Answer: C

27. All of the following are considered a kind of contraction EXCEPT
    A) tetanus.
    B) twitch.
    C) isotonic.
    D) isometric.
    E) All of the above are contractions.
    Answer: E

28. The attachment of a muscle to the stationary bone is called
    A) insertion.
    B) fixation.
    C) head.
    D) origin.
    E) ligation.
    Answer: D

29. Which of the following is NOT a characteristic used to name skeletal muscles?
    A) direction of fibers.
    B) thickness of fibers.
    C) location.
    D) size.
    E) number of origins.
    Answer: B

30. A muscle that has three origins is called a
    A) bicep.
    B) tricep.
    C) quadricep.
    D) deltoid.
    E) trapezius.
    Answer: B

31. A muscle that produces an upward movement is a
    A) pronator.
    B) tensor.
    C) flexor.
    D) depressor.
    E) levator.
    Answer: E

32. A muscle that decreases the size of an opening is a
    A) rotator.
    B) tensor.
    C) pronator.
    D) sphincter.
    E) depressor.
    Answer: D

33. Which of the following is a muscle of the facial expression?
    A) eternal oblique
    B) rhomboideus
    C) teres major
    D) platysma
    E) pronator teres
    Answer: D

34. The muscle that produces the action of sucking is the
    A) buccinator.
    B) zygomaticus.
    C) frontalis.
    D) epicranius.
    E) occipitalis.
    Answer: A

35. Which of the following is a muscle that moves the mandible?
    A) orbicularis occuli.
    B) masseter.
    C) orbicularis oris.
    D) epicranius.
    E) frontalis.
    Answer: B

36. All of the following are muscles that move the eyeball EXCEPT
    A) superior rectus.
    B) lateral rectus.
    C) inferior oblique.
    D) internal oblique.
    E) superior oblique.
    Answer: D

37. Which of the following is a muscle that moves the pectoral girdle?
    A) trapezius
    B) pectoralis
    C) latissimus dorsi
    D) gracillus
    E) sartorius
    Answer: A

38. All of the following are muscles that move the humerus EXCEPT
    A) latissimus dorsi.
    B) infraspinatus.
    C) teres major.
    D) deltoid.
    E) pronator teres.
    Answer: E

39. Which of the following flexes and abducts the wrist?
    A) flexor digitorum profundus
    B) flexor digitorum superficialis
    C) flexor carpi radialis
    D) flexor carpi ulnaris
    E) palmaris longus
    Answer: C

40. The sternocleidomastoid is a muscle that moves the
    A) fingers.
    B) mandible.
    C) maxilla.
    D) humerus.
    E) vertebral column.
    Answer: E

41. All of the following are muscles of the lower limb EXCEPT
    A) extensor digitorum.
    B) adductor magnus.
    C) adductor longus.
    D) quadriceps femoris.
    E) vastus medialis.
    Answer: A

42. Which of the following is a muscle that moves the foot?
    A) peroneus magnus
    B) semitendinosus
    C) sartorius
    D) biceps femoris
    E) gracilis
    Answer: A

43. Which of the following muscles adducts the thigh and flexes the leg?
    A) rectus femoris
    B) gracilis
    C) tibialis anterior
    D) gastronemicus
    E) flexor digitorum longus
    Answer: B

44. The infraspinatus muscle is a muscle that moves the
    A) upper leg.
    B) lower leg.
    C) upper arm.
    D) lower arm.
    E) carpals.
    Answer: C

45. Pain in or associated with muscles is called
    A) hypertonia.
    B) proprioception.
    C) myotonia.
    D) myalgia.
    E) myositis.
    Answer: D

## True-False

*Write T if the statement is true and F if the statement is false.*

1. Skeletal muscle tissue is striated and voluntary.
   Answer: True

2. One byproduct of skeletal muscle contraction is heat.
   Answer: True

3. The myofibrils of skeletal muscle fibers are composed of three types of myofilaments.
   Answer: False

4. The H zone is located in the center of the I band.
   Answer: False

5. During muscle contraction, the myosin heads pull on the actin filaments, causing the actin to slide toward the center of the sarcomere.
   Answer: True

6. The ends of axon terminals enlarge into swellings called motor end plates.
   Answer: False

7. The power stroke in muscle contraction is the force causing the actin filaments to slide between the myosin filaments.
   Answer: True

8. The detachment of myosin heads from actin requires ATP.
   Answer: True

9. Any stimulus from a neuron can initiate a muscle contraction.
   Answer: False

10. Muscle tissue plays a major role in maintenance of homeostasis.
    Answer: True

11. Skeletal muscle has a long refractory period.
    Answer: False

12. When a muscle shortens and pulls on another structure, it is undergoing an isotonic contraction.
    Answer: True

13. Muscle tone requires the contraction of an entire muscle.
    Answer: False

14. There are two types of smooth muscle tissue, visceral and multiunit.
    Answer: True

15. Smooth muscle is under voluntary control.
    Answer: False


## Short Answer


*Write the word or phrase that best completes each statement or answers the question.*

1. The sheet of fibrous connective tissue around muscles is _____.
   Answer: fascia

2. The electrical current that travels over the sarcolemma and along the transverse tubules is called ____.
   Answer: muscle action potential

3. The cytoplasm of a muscle fiber is called ____.
   Answer: sarcoplasm

4. The basic, functional units of skeletal muscle fibers are ____.
   Answer: sarcomeres

5. During muscle contraction filaments ____, they do not shorten.
   Answer: slide

6. A motor neuron and the muscle fibers it stimulates is called a ____.
   Answer: motor unit

7. Synaptic vesicles are filled with chemicals called ____.
   Answer: neurotransmitter

8. The ion absolutely essential for the sliding of the filaments in muscle contraction is ____.
   Answer: calcium

9. A poison used by South American Indians to cause muscle paralysis is _____.
   Answer: curare

10. _____ is a high energy molecule in skeletal muscle fibers, which can be used to produce ATP as needed.
    Answer: creatine phosphate

11. In order for a muscle fiber to contract, a _____ stimulus has to be applied.
    Answer: threshold

12. In response to a stimulus, muscle fibers either contract or do not contract, which is in accordance with the _____ principle.
    Answer: all-or-none

13. The inability of a muscle to maintain its strength is called _____.
    Answer: (muscle) fatigue

14. The period of time between the application of a stimulus and the beginning of muscle contraction is called _____.
    Answer: latent period

15. A contraction where the tension of the muscle increases, but the shortening is minimal, is called a(n) _____ contraction.
    Answer: isometric

16. The intermediate filaments in smooth muscle fibers stretch between structures called _____.
    Answer: dense bodies

17. A muscle that causes a desired action is referred to as the _____.
    Answer: prime mover (agonist)

18. Derivatives of testosterone that are illegally used to build up muscle proteins are called _____ steroids.
    Answer: anabolic

19. The muscle that moves a bone around its longitudinal axis is called a _____.
    Answer: rotator

20. The maximum ability to move bones about the joint through an arc of a circle is a joint's
    _____ _____ _____.
    Answer: range of motion

21. The autoimmune disorder involving the neuromuscular junction is _____.
    Answer: myasthenia gravis

22. At death, when ATP is not available for the detachment of the myosin head from the actin, _____ occurs.
    Answer: rigor mortis

23. Epimysium, perimysium, and endomysium extend beyond the muscle as a _____.
    Answer: tendon

24. A better term for oxygen debt is _____ _____ _____.
    Answer: recovery oxygen uptake

25. The process in which the number of contracting motor units is increased is called _____ _____
    _____.
    Answer: motor unit recruitment

## Matching

*Choose the item from Column 2 that best matches each item in Column 1.*

1.  Column 1: Nonstriated, involuntary muscle
    Column 2: smooth muscle

2.  Column 1: Striated involuntary muscle
    Column 2: cardiac muscle

3.  Column 1: Consists of mostly thick filaments
    Column 2: A band

4.  Column 1: Consists of mostly thin filaments
    Column 2: I band

5.  Column 1: It is located in the center of each A band
    Column 2: H zone

6.  Column 1: A type of neuron that stimulates muscle tissue
    Column 2: motor neuron

7.  Column 1: The region of the sarcolemma near the axon terminal
    Column 2: motor end plate

8.  Column 1: A process that does not require oxygen
    Column 2: anaerobic respiration

9.  Column 1: A brief contraction of muscle fibers
    Column 2: Twitch

10. Column 1: Continued, sustained smooth contraction due to rapid stimulation
    Column 2: tetanus

## Essay

*Write your answer in the space provided or on a separate sheet of paper.*

1.  Name and describe the three types of muscle tissue.
    Answer: Skeletal muscle tissue is striated and under voluntary control. Cardiac muscle tissue
    is also striated but is involuntary. Cardiac muscle fibers are connected by intercalated
    discs. Smooth muscle fibers are nonstriated and involuntary.

2.  Describe the connective tissue components of skeletal muscle tissue.
    Answer: An entire muscle is surrounded by a connective tissue covering called epimysium, the
    individual fascicles of a muscle are surrounded by perimysium, and individual muscle fibers
    are surrounded by endomysium.

3.  Define muscle tone.
    Answer: A sustained partial contraction of portions of skeletal muscle results in muscle tone, which is
    essential for the maintenance of posture.

## CHAPTER 9   Nervous Tissue

## Multiple-Choice

*Choose the one alternative that best completes the statement or answers the question.*

1. The science that deals with the normal functioning and the disorders of the nervous system is called
   A) pathology.
   B) endocrinology.
   C) neuroscience.
   D) neurology.
   E) neuropathology.
   Answer: D

2. The nervous system and the _____ system share the greatest responsibility for maintaining homeostasis.
   A) immune.
   B) endocrine.
   C) cardiovascular.
   D) respiratory.
   E) sensory.
   Answer: B

3. Neurons that conduct nerve impulses from the receptors to the central nervous system are
   A) motor neurons.
   B) efferent neurons.
   C) interneurons.
   D) sensory neurons.
   E) association neurons.
   Answer: D

4. Processes that carry impulses to another neuron are
   A) dendrites.
   B) axons.
   C) synapses.
   D) axon collaterals.
   E) myelin sheaths.
   Answer: B

5. Which of the following is an example of an effector?
   A) interneuron
   B) thermoreceptor
   C) proprioceptor
   D) nerve
   E) glands
   Answer: E

6. All of the following are part of the PNS EXCEPT:
   A) neuroendocrine system.
   B) somatic nervous system.
   C) autonomic nervous system.
   D) enteric nervous system.
   E) sympathetic branch.
   Answer: A

7. Enteric motor neurons are responsible for
   A) contraction of skeletal muscle close to the GI tract.
   B) contraction of GI tract smooth muscle.
   C) contraction of smooth muscle of the blood vessels.
   D) Both A and B.
   E) Both B and C.
   Answer: B

8. The neuroglia that produce myelin in the peripheral nervous system are
   A) neurolemmocytes.
   B) oligodendrocytes.
   C) microglia.
   D) astrocytes.
   E) pituicytes.
   Answer: A

9. A group of nerve fibers in the central nervous system is called a(n)
   A) axon.
   B) nerve.
   C) tract.
   D) ganglion.
   E) nucleus.
   Answer: C

10. Groups of neuron cell bodies in the peripheral nervous system are called
    A) ganglia.
    B) nuclei.
    C) horns.
    D) tracts.
    E) nerves.
    Answer: A

11. A disorder of the central nervous system would involve
    A) sympathetic neurons.
    B) senses.
    C) brain.
    D) spinal cord.
    E) brain and spinal cord.
    Answer: E

12. Which of the following cells does NOT functionally belong with the others?
    A) efferent neurons
    B) oligodendrocytes
    C) afferent neurons
    D) motor neuron
    E) interneuron
    Answer: B

13. The portion of the nervous system that is considered involuntary is the
    A) somatic nervous system.
    B) sensory nervous system.
    C) autonomic nervous system.
    D) motor nervous system.
    E) peripheral nervous system.
    Answer: C

14. The portion where an axon joins the cell body is the
    A) initial segment.
    B) axon hillock.
    C) myelin sheath.
    D) neurolemma.
    E) axon collateral.
    Answer: B

15. In the spinal cord, the gray matter is located in regions called
    A) tracts.
    B) horns.
    C) nuclei.
    D) column.
    E) ganglia.
    Answer: B

16. All of the following are functions of the nervous system EXCEPT
    A) senses changes.
    B) analyzes changes.
    C) stores potassium.
    D) responses to changes.
    E) integrates impulses.
    Answer: C

17. Compared to unmyleninated axons, myelinated axons
    A) conduct impulses faster.
    B) conduct impulses more often.
    C) conduct impulses slower.
    D) produce larger action potentials.
    E) produce longer lasting action potentials.
    Answer: A

18. The different charge between the outside and the inside of a neuron at rest is called
    A) action potential.
    B) synaptic potential.
    C) excitatory postsynaptic potential.
    D) resting membrane potential.
    E) equilibrium potential.
    Answer: D

19. The ability of nerve cells to respond to stimuli and to convert them into nerve impulses is called
    A) depolarization.
    B) hyperpolarization.
    C) excitability.
    D) threshold.
    E) contractility.
    Answer: C

20. The neuroglial cell involved in the blood brain barrier is the
    A) astrocyte.
    B) oligodendrocyte.
    C) microglia.
    D) satellite glia.
    E) ependymal cell.
    Answer: A

21. Saltatory conduction refers to the conduction of impulses in
    A) cardiac muscle.
    B) skeletal muscle.
    C) unmyelinated fibers.
    D) myelinated fibers.
    E) all axons.
    Answer: D

22. Depolarization of a membrane is due to the
    A) opening of chlorine channels.
    B) opening of sodium channels.
    C) opening of potassium channels.
    D) closing of potassium channels.
    E) closing of sodium channels.
    Answer: B

23. The stage in an action potential that immediately follows depolarization is
    A) hyperpolarization.
    B) polarization.
    C) repolarization.
    D) threshold.
    E) the resting period.
    Answer: C

24. In order for an action potential to occur the cell membrane
    A) has to be hyperpolarized.
    B) reach the repolarization phase.
    C) reach the refractory period.
    D) reach threshold.
    E) All of the above.
    Answer: D

25. The period of time during which the neuron cannot generate another action potential is called
    A) hyperpolarization.
    B) threshold period.
    C) repolarization.
    D) refractory period.
    E) depolarization.
    Answer: D

26. A Stimulus strong enough to generate a nerve impulse is called
    A) threshold stimulus.
    B) action potential.
    C) subthreshold stimulus.
    D) conducting stimulus.
    E) Both A and D are correct.
    Answer: A

27. The speed of nerve impulse conduction is determined by all of the following EXCEPT
    A) temperature.
    B) fiber diameter.
    C) presence of myelin.
    D) absence of myelin.
    E) stimulus strength.
    Answer: E

28. The junction between two nerve cells is called
    A) neuromuscular junction.
    B) neuroglandular junction.
    C) gap junction.
    D) synapse.
    E) synaptic terminal.
    Answer: D

29. The special mode of impulse travel is called
    A) propagation.
    B) conduction.
    C) refraction.
    D) A and B.
    E) A and C.
    Answer: D

30. Neurotransmitters are stored in
    A) the synaptic cleft.
    B) synaptic vesicles.
    C) the postsynaptic neuron.
    D) Nissl bodies.
    E) smooth endoplasmic reticulum.
    Answer: B

31. Which of the following is necessary for the release of the neurotransmitter into the synaptic cleft?
    A) sodium
    B) iron
    C) calcium
    D) potassium
    E) magnesium
    Answer: C

32. What is the result of inhibitory neurotransmission on the postsynaptic membrane?
    A) a generator potential
    B) an action potential
    C) depolarization
    D) hyperpolarization
    E) excitation
    Answer: D

33. Neurotransmitters are released at the
    A) dendrite.
    B) axon collateral.
    C) axon terminal.
    D) axon hillock.
    E) cell body.
    Answer: C

34. A stimulus weaker than a stimulus resulting in a nerve impulse is a(n)
    A) light stimulus.
    B) hyperpolarizing stimulus.
    C) threshold stimulus.
    D) subthreshold stimulus.
    E) depolarizing stimulus.
    Answer: D

35. Schwann cells are also called
    A) neurolemmocytes.
    B) astrocytes.
    C) oligodendrocytes.
    D) microglia.
    E) satellite cells.
    Answer: A

36. Which of the following provides myelin for axons in the central nervous system?
    A) Schwann cells
    B) neurolemmocytes
    C) astrocytes
    D) microglia
    E) oligodendrocytes
    Answer: E

37. Ascending tracts
    A) are found in the spinal cord.
    B) carry sensory impulses.
    C) carry motor impulses.
    D) Both A and B.
    E) Both A and C.
    Answer: D

38. Nerve fiber is the term for
    A) axons only.
    B) dendrites only.
    C) an axon or a dendrite.
    D) myelinated axons only.
    E) unmyelinated axons only.
    Answer: C

39. Most neurons in the central nervous system are
    A) sensory neurons.
    B) interneurons.
    C) motor neurons.
    D) efferent neurons.
    E) afferent neurons.
    Answer: B

40. Processes that receive impulses and conduct them toward the cell body are
    A) dendrites.
    B) axons.
    C) axon collaterals.
    D) synaptic end bulbs.
    E) neurolemmocytes.
    Answer: A

41. Clusters of neuronal cell bodies and dendrites in the brain are called
    A) ganglia.
    B) tracts.
    C) regions.
    D) nuclei.
    E) columns.
    Answer: D

42. Bundles of unmyelinated axons, dendrites, and neuron cell bodies form
    A) nuclei.
    B) white matter.
    C) gray matter.
    D) columns.
    E) tracts.
    Answer: C

43. All of the following statements about ion channels are true EXCEPT
    A) they are present in plasma membranes.
    B) they are formed by membrane proteins.
    C) some channels are always open.
    D) some open and close in response to chemicals.
    E) all channels respond to changes in voltage.
    Answer: E

44. The duration of a nerve impulse is approximately
    A) one second.
    B) a tenth of a second.
    C) a microsecond.
    D) a millisecond.
    E) a fifth of a second.
    Answer: D

45. Which of the following is an inhibitory neurotransmitter?
    A) glutamate
    B) glycine
    C) endorphin
    D) acetylcholine
    E) serotonin
    Answer: B


## True-False

*Write T if the statement is true and F if the statement is false.*

1. The branch of medicine that deals with the functioning and disorders of the nervous system is called neurophysiology
   Answer: false

2. The cranial nerves belong to the peripheral nervous system.
   Answer: True

3. Ependymal cells are neuroglial cells.
   Answer: True

4. In general, neurons in adults cannot regenerate.
   Answer: True

5. Repolarization is due to the opening of sodium channels.
   Answer: False

6. If the resting membrane potential becomes more negative the membrane is depolarized.
   Answer: False

7. Another term for nerve impulse is action potential.
   Answer: True

8. Muscle action potentials are similar to nerve action potentials.
   Answer: True

9. Continuous conduction occurs along myelinated axons.
   Answer: False

10. When nerve fibers are cooled, impulse conduction occurs at higher speeds.
    Answer: False

11. The junction between a neuron and a glandular cell is called a neuroglandular junction.
    Answer: True

12. The postsynaptic neuron is considered to be an integrator.
    Answer: True

13. Some neural regeneration can occur in the peripheral nervous system.
    Answer: True

14. Subthreshold stimulation can cause action potentials.
    Answer: False

15. Neurotransmitters always cause hyperpolarization of the synaptic membrane.
    Answer: False


## Short Answer

*Write the word or phrase that best completes each statement or answers the question.*

1. The nervous system has two principal divisions, the ____ and the ____.
   Answer: central nervous system, peripheral nervous system

2. Nerve cells that conduct impulses from the central nervous system to muscles and glands are ____ neurons.
   Answer: motor or efferent

3. Side branches of axons are called ____.
   Answer: axon collaterals

4. Sacs that store chemicals in the axon terminals area ____.
   Answer: synaptic vesicles

5. A covering around axons, which is produced by neuroglia, is called ____.
   Answer: myelin sheath

6. Nerves that contain both sensory and motor fibers are _____ nerves.
   Answer: mixed

7. The network of neurons in the walls of the organs of the gastrointestinal tract regulating the digestive system is the ____ _____.
   Answer: enteric plexus

8. The gaps in the myelin sheath along axons are called _____.
   Answer: neurofibral nodes (nodes of Ranvier)

9. Spinal tracts that carry motor impulses down the cord are classified as _____ tracts.
   Answer: descending

10. The sodium-potassium pump is a(n) _____ transport mechanism.
    Answer: active

11. The different charge on the inside and outside of a resting neuron is called _____.
    Answer: resting membrane potential

12. The critical level of depolarization that has to be reached for an action potential to occur is the _____.
    Answer: threshold

13. When the polarity of a membrane becomes more negative than the resting membrane potential, the membrane is considered to be _____.
    Answer: hyperpolarized

14. The type of impulse conduction in myelinated fibers is called _____.
    Answer: saltatory

15. The connection between a neuron and a muscle fiber it innervates is called _____.
    Answer: neuromuscular junction

16. The neuron before a synapse is a _____ neuron, the neuron located after a synapse is a _____.
    Answer: presynaptic, postsynaptic

17. Axon terminals end in bulb-like structures called _____.
    Answer: synaptic end bulbs

18. The capacity of cells to repair or replicate themselves is called _____.
    Answer: regeneration

19. Large diameter fibers conduct impulses _____ than small diameter fibers.
    Answer: faster

20. The chemicals stored in synaptic vesicles are referred to as _____.
    Answer: neurotransmitters

21. The space between a presynaptic and postsynaptic neuron is called _____.
    Answer: synaptic cleft

22. A nerve cell transmits action potentials according to the _____ principle.
    Answer: all-or-none

23. Channels that open in response to a change in the membrane potential are _____ _____

    _____.
    Answer: voltage-regulated channels

24. Synapses which conduct action potentials directly from the presynaptic to the postsynaptic cell are _____ synapses.
    Answer: electrical

25. A progressive destruction of the myelin sheaths of neurons in the CNS is _____ _____.
    Answer: multiple sclerosis

## Matching

*Choose the item from Column 2 that best matches each item in Column 1.*

1.  Column 1: The time during which the neuron cannot generate another action potential.
    Column 2: refractory period

2.  Column 1: Also called a nerve impulse.
    Column 2: action potential

3.  Column 1: The loss of polarization
    Column 2: depolarization

4.  Column 1: An increase in the negativity of the membrane potential
    Column 2: hyperpolarization

5.  Column 1: The ability of a cell to respond to stimuli and to convert them into nerve impulses.
    Column 2: excitability

6.  Column 1:  An insufficient stimulus
    Column 2: subthreshold stimulus

7.  Column 1: Collection of nerve fibers in the CNS.
    Column 2: tracts

8.  Column 1: A collection of nerve cell bodies in the PNS.
    Column 2: ganglia

9.  Column 1: Produces myelin in the PNS.
    Column 2: neurolemmocyte

10. Column 1: Carries nerve impulses away from the cell body.
    Column 2: axon

11. Column 1: Carries impulses to the cell body.
    Column 2: dendrite

## Essay

*Write your answer in the space provided or on a separate sheet of paper.*

1.  Name and describe the function of the different cytoplasmic processes in neurons.
    Answer: Axons carry impulses away from the cell body to another neuron or tissue, while
    dendrites carry impulses to the cell body.

2.  Describe nerve impulse conduction in unmyelinated and myelinated axons.
    Answer: In unmyelinated axons, impulses travel by continuous conduction which means that the
    action potential in one area causes depolarization of the adjacent area. This step-by-step
    depolarization is slower conduction than occurs in myelinated axons. In myelinated axons,
    depolarization occurs at the nodes of Ranvier and impulses jump from one node to the next.
    This conduction is called saltatory conduction.

3.  Name the six types of neuroglia and give their general function.
    Answer:  Astrocytes form part of the blood brain barrier, oligodendrocytes provide the myelin
    in the CNS, neuroglia are phagocytotic cells in the CNS, ependymal cells line the
    brain ventricles, neurolemmocytes (Schwann cells) provide the myelin in the PNS, and
    satellite cells are supportive glia in the PNS.

## CHAPTER 10   Central and Somatic Nervous Systems

## Multiple-Choice

*Choose the one alternative that best completes the statement or answers the question.*

1.  The subarachnoid space is located between the
    A)  arachnoid and pia mater.
    B)  arachnoid and dura mater.
    C)  dura mater and pia mater.
    D)  bone and pia mater.
    E)  bone and arachnoid.
    Answer: A

2.  The outermost of the meninges is the
    A)  pia mater.
    B)  arachnoid.
    C)  dura mater.
    D)  epidural layer.
    E)  subdural layer.
    Answer: D

3.  The dura mater is composed of
    A)  smooth muscle.
    B)  adipose tissue.
    C)  loose connective tissue.
    D)  dense regular connective tissue.
    E)  dense irregular connective tissue.
    Answer: E

4.  The cervical enlargement contains nerves that supply the
    A)  upper limbs.
    B)  lower limbs.
    C)  thorax.
    D)  vertebral column.
    E)  None of the above.
    Answer: A

5.  The axillary nerve belongs to the
    A)  cervical plexus.
    B)  brachial plexus.
    C)  intercostal nerves.
    D)  lumbar plexus.
    E)  sacral plexus.
    Answer: B

6.  Which of the following is NOT a term used to describe a spinal cord structure?
    A)  central canal
    B)  anterior median fissure
    C)  corpus callosum
    D)  lumbar enlargement
    E)  cauda equina
    Answer: C

7.  All of the following is information traveling in the fasciculus gracilis and fasciculus cuneatus EXCEPT
    A) proprioception.
    B) pain.
    C) vibratory sensation.
    D) weight discrimination.
    E) stereognosis.
    Answer: B

8.  The dorsal root contains
    A) sensory fibers only.
    B) motor fibers only.
    C) both sensory and motor fibers.
    D) autonomic fibers only.
    E) somatic fibers only.
    Answer: A

9.  In the reflex arc, a muscle or gland is considered to be the
    A) receptor.
    B) integrating center.
    C) sensor.
    D) motor neuron.
    E) effector.
    Answer: E

10. All of the following are autonomic reflexes EXCEPT
    A) withdrawal reflex.
    B) coughing.
    C) sneezing.
    D) swallowing.
    E) defecation.
    Answer: A

11. Autonomic reflexes involve
    A) skeletal muscle.
    B) smooth muscle.
    C) cardiac muscle.
    D) Both A and B.
    E) Both B and C.
    Answer: E

12. Which of the branches of the spinal nerves supplies the deep muscles and the skin of the back?
    A) anterior ramus
    B) posterior ramus
    C) cervical plexus
    D) lumbar plexus
    E) brachial plexus
    Answer: B

13. Which of the spinal nerves do NOT form a plexus?
    A) C1-C5
    B) C5-T1
    C) T2-T11
    D) T12-L4
    E) L4-S4
    Answer: C

14. Which spinal cord plexus supplies the abdominal wall?
    A) cervical plexus
    B) sacral plexus
    C) brachial plexus
    D) lumbar plexus
    E) brachial and sacral plexuses
    Answer: D

15. Damage to the choroid plexus would interfere with
    A) myelin production.
    B) formation of cerebrospinal fluid.
    C) glycogen storage.
    D) motor movements.
    E) impulse conduction.
    Answer: B

16. The portion of the brain which is continuous with the spinal cord is the
    A) cerebrum.
    B) pons.
    C) medulla.
    D) midbrain.
    E) cerebellum.
    Answer: C

17. The cardiovascular center is located in the
    A) hippocampus.
    B) cerebrum.
    C) diencephalon.
    D) medulla oblongata.
    E) insula.
    Answer: D

18. The thalamus is part of the
    A) brain stem.
    B) diencephalon.
    C) cerebrum.
    D) cerebellum.
    E) hypothalamus.
    Answer: B

19. The nucleus of cranial nerve number VII (facial nerve) is located in the
    A) pons.
    B) medulla.
    C) midbrain.
    D) cerebellum.
    E) cerebrum.
    Answer: A

20. The cerebral peduncles are part of the
    A) pons.
    B) medulla.
    C) midbrain.
    D) insula.
    E) hypothalamus.
    Answer: C

21. Which of the following areas deals with emotion?
    A) reticular activating system
    B) thalamus
    C) brain stem
    D) limbic system
    E) cerebrum
    Answer: D

22. Which of the areas of the cerebral cortex receives sensation from cutaneous and muscular receptors?
    A) association areas
    B) primary somatosensory area
    C) primary motor area
    D) gnostic area
    E) premotor area
    Answer: B

23. Which of the following is NOT a ventricle in an adult brain?
    A) lateral ventricle
    B) first ventricle
    C) third ventricle
    D) fourth ventricle
    E) All of the above are ventricles in the adult brain.
    Answer: B

24. The primary visual area is located in the
    A) occipital lobe.
    B) temporal lobe.
    C) frontal lobe.
    D) Broca's area.
    E) lateral lobe.
    Answer: A

25. All of the following areas are associated with memory EXCEPT
    A) parts of the limbic system.
    B) parts of the diencephalon.
    C) frontal lobe.
    D) parietal, occipital, and temporal lobes.
    E) cerebellum.
    Answer: E

26. The neurotransmitter which stimulates perception of pain is
    A) acetylcholine.
    B) substance P.
    C) endorphin.
    D) dopamine.
    E) GABA.
    Answer: B

27. The progressive destruction of myelin sheaths in the central nervous system is called
    A) multiple sclerosis.
    B) Parkinson's disease.
    C) dyslexia.
    D) epilepsy.
    E) Alzheimer's disease.
    Answer: A

28. Which of the following numbers belongs to the trochlear nerve?
    A) I
    B) III
    C) IV
    D) VII
    E) X
    Answer: C

29. Which structure does NOT belong with the others?
    A) medulla
    B) pons
    C) brain stem
    D) thalamus
    E) midbrain
    Answer: D

30. Which of the following cranial nerves is sensory only?
    A) olfactory
    B) oculomotor
    C) facial
    D) glossopharyngeal
    E) hypoglossal
    Answer: A

31. The cranial nerve that innervates the gastrointestional tract is the
    A) hypoglossal (XII).
    B) accessory (XI).
    C) vagus (X).
    D) abducens (VI).
    E) None of the cranial nerves innervate the gastrointestinal tract.
    Answer: C

32. The area of the brain that regulates posture and balance is the
    A) limbic system.
    B) cerebrum.
    C) thalamus.
    D) cerebellum.
    E) brain stem.
    Answer: D

33. In a spinal tap, cerebrospinal fluid is removed from the
    A) epidural space.
    B) foramen magnum.
    C) subdural space.
    D) fourth ventricle.
    E) subarachnoid space.
    Answer: E

34. All of the following are part of the reflex arc EXCEPT
    A) sensory neuron.
    B) motor neuron.
    C) hormones.
    D) effectors.
    E) nerve impulses.
    Answer: C

35. The intercostal nerves innervate the
    A) lower limbs.
    B) muscles between the ribs.
    C) muscles of the upper arm.
    D) lungs.
    E) shoulder.
    Answer: B

36. The pyramids of the brain belong to the
    A) diencephalon.
    B) cerebrum.
    C) cerebellum.
    D) medulla oblongata.
    E) hypothalamus.
    Answer: D

37. The IIIrd and IVth cranial nerves originate in the
    A) reticular formation.
    B) thalamus.
    C) midbrain.
    D) pons.
    E) medulla.
    Answer: C

38. A shallow groove on the surface of the cortex is a
    A) fissure.
    B) sulcus.
    C) gyrus.
    D) furrow.
    E) indentation.
    Answer: B

39. All of the following are functions of the hypothalamus EXCEPT
    A) control of the pituitary gland.
    B) regulation of eating.
    C) regulation of respiration.
    D) regulation of body temperature.
    E) satiety center.
    Answer: C

40. Which of the following is NOT a basal ganglion?
    A) reticular formation
    B) corpus striatum
    C) putamen
    D) lentiform nucleus
    E) caudate nucleus
    Answer: A

41. Parkinson's disease is due to the degeneration of dopamine-producing neurons in the
    A) cerebellum.
    B) substantia nigra.
    C) medulla.
    D) hippocampus.
    E) cerebral peduncles.
    Answer: B

42. The term for pain relief is
    A) analgesia.
    B) agnosia.
    C) anethesia.
    D) dementia.
    E) nerve block.
    Answer: A

43. The primary auditory area is located in the
    A) occipital lobe.
    B) postcentral gyrus.
    C) frontal lobe.
    D) temporal lobe.
    E) parietal lobe.
    Answer: D

44. Which of the cranial nerves is the optic nerve?
    A) I
    B) II
    C) III
    D) VI
    E) V
    Answer: B

45. The herpes zoster virus causes
    A) meningitis.
    B) neuralgia.
    C) shingles.
    D) agnosia.
    E) encephalitis.
    Answer: C


## True-False

*Write T if the statement is true and F if the statement is false.*

1. The middle layer of the meninges is called pia mater.
   Answer: False

2. The cervical enlargement contains nerves that supply the upper extremities.
   Answer: True

3. Pyramidal pathways convey nerve impulses that program autonomic movements.
   Answer: False

4. The dorsal root ganglion contains cell bodies of sensory neurons.
   Answer: True

5. The withdrawal reflex is an autonomic reflex.
   Answer: False

6. Spinal nerves are mixed nerves.
   Answer: True

7. The brain contains approximately 100 million neurons.
   Answer: False

8. There are four lateral ventricles in the brain.
   Answer: False

9. The supply of glucose to the brain must be continuous.
   Answer: True

10. The crossing of fibers to the opposite site of the spinal cord is called decussation.
    Answer: True

11. The satiety center is located in the medulla oblongata.
    Answer: False

12. The central sulcus separates the temporal and parietal lobes.
    Answer: False

13. The largest basal ganglion is the corpus striatum.
    Answer: True

14. The auditory association area is located in the diencephalon.
    Answer: False

15. Dopamine is an excitatory neurotransmitter.
    Answer: True

## Short Answer

*Write the word or phrase that best completes each statement or answers the question.*

1. An injection of an anesthetic into the epidural space results in a(n) _____.
   Answer: epidural block

2. An inflammation of the meninges is known as _____.
   Answer: meningitis

3. The axillary nerve originates from the _____ plexus.
   Answer: brachial

4. The patellar reflex is a(n) _____ reflex.
   Answer: somatic

5. Spinal nerves divide into several branches, known as _____.
   Answer: rami

6. The sciatic nerve arises from the _____ plexus.
   Answer: sacral

7. The spinothalamic tract is a(n) _____ tract.
   Answer: ascending

8. The fluid that circulates through the ventricles is the _____.
   Answer: cerebrospinal fluid (CSF)

9. The circulatory route at the base of the brain is called the _____.
   Answer: cerebral arterial circle or circle of Willis

10. The barrier between the blood and the brain is called _____.

11. Humans awaken and sleep in a fairly constant rhythm called a ____ rhythm.
    Answer: circadian

12. The primary olfactory area is located in the ____ lobe.
    Answer: temporal

13. Memory that lasts for only a few seconds is ____.
    Answer: short-term memory

14. Brain waves can be measured with the ____.
    Answer: electroencephalogram (EEG)

15. Nerves arising from the lowest portion of the spinal cord are called the ____.
    Answer: cauda equina

16. The center for thirst is located in the ____.
    Answer: hypothalamus

17. The roughly triangular structures in the medulla that carry motor impulses, are called ____.
    Answer: pyramids

18. The CSF has two functions related to homeostasis: protection and ____.
    Answer: circulation

19. The vagus nerve originates in the ____.
    Answer: medulla

20. A group of widely scattered neurons with a net-like appearance is referred to as ____.
    Answer: reticular formation

21. The gray matter of the cerebrum is found in the ____.
    Answer: cerebral cortex

22. The lenticular nucleus is subdivided into the globus pallidus and the ____.
    Answer: putamen

23. The pons is part of the ____ ____.
    Answer: brain stem

24. Cerebrospinal fluid is produced by the ____ ____.
    Answer: choroid plexus

25. Inactivation of the RAS produces ____.
    Answer: sleep

## Matching

*Choose the item from Column 2 that best matches each item in Column 1.*

1. Column 1: Optic nerve
   Column 2: II

2. Column 1: Trochlear nerve
   Column 2: IV

3. Column 1: Abducens
   Column 2: VI

4. Column 1: Glossopharyngeal
   Column 2: IX

5. Column 1: Accessory
   Column 2: XI

6. Column 1: A cerebrovascular accident.
   Column 2: Stroke

7. Column 1: A disabling memory disorder.
   Column 2: Alzheimer's disease

8. Column 1: Paralysis of both lower limbs.
   Column 2: Paraplegia

9. Column 1: A group of motor disorders resulting in the loss of muscle contraction.
   Column 2: Cerebral palsy

10. Column 1: A disorder that involves dopamine.
    Column 2: Parkinson's disease

## Essay

*Write your answer in the space provided or on a separate sheet of paper.*

1. Name and briefly describe the meninges.
   Answer: The dura mater is the outermost layer and consists of tough, dense, irregular connective tissue. The arachnoid is the middle layer arranged in delicate webs of collagen and elastic fibers. The inner layer is the pia mater, a transparent layer of collagen and elastic fibers.

2. Name the principal parts of the brain and their subdivisions.
   Answer: Cerebrum: cerebral cortex and cerebral white matter
   Diencephalon: thalamus and hypothalamus
   Brain stem: medulla, pons, and midbrain
   Cerebellum

3. Name the brain ventricles and the structures which connect them.
   Answer: The two lateral ventricles are connected to the third ventricle by the interventricular foramen. The third ventricle is connected to the fourth ventricle by the cerebral aqueduct, and the fourth ventricle is continuous with the central canal of the spinal cord.

4. Describe the functions of the hypothalamus.
   Answer: The hypothalamus controls and integrates autonomic system activities, it controls the release of many hormones of the pituitary gland, it controls body temperature, and it regulates food intake. The hypothalamus is also associated with feelings such as aggression, pain, and pleasure. The thirst center is also found in the hypothalamus. Furthermore, the hypothalamus is one of the areas that maintains consciousness and sleep patterns.

# CHAPTER 11    Autonomic Nervous System

## Multiple-Choice

*Choose the one alternative that best completes the statement or answers the question.*

1. Which of the following statements about the autonomic nervous system is NOT true?
   A) it regulates the size of the pupils
   B) all of the fibers are afferent fibers
   C) it consists of two subdivisions
   D) it regulates visceral activities
   D) it is an involuntary system
   Answer:  B

2. The first motor neuron in an autonomic pathway is called a
   A) preganglionic neuron.
   B) postganglionic neuron.
   C) autonomic ganglion.
   D) sympathetic neuron.
   E) parasympathetic neuron.
   Answer:  A

3. The part of the nervous system that regulates smooth muscle is the
   A) somatic nervous system.
   B) limbic system.
   C) autonomic nervous system.
   D) Both A and B.
   E) Both B and C.
   Answer: C

4. Which of the following is an example of an input to the ANS?
   A) temperature
   B) blood carbon dioxide levels
   C) proprioception
   D) equilibrium
   E) both C and D
   Answer: B

5. The sympathetic division of the ANS is also referred to as the
   A) autonomic division.
   B) craniosacral division.
   C) thoracolumbar division.
   D) somatic division.
   E) ganglionic division.
   Answer:  C

6. All of the following are autonomic ganglia EXCEPT
   A) sympathetic trunk ganglia.
   B) prevertebral ganglia.
   C) terminal ganglia.
   D) dorsal root ganglia.
   E) All of the above are autonomic ganglia.
   Answer: D

7. Which of the following statements does NOT apply to the autonomic nervous system?
   A) Autonomic pathways consist of sets of two motor neurons.
   B) ANS neurons always release acetylcholine as their neurotransmitter.
   C) The output part of the ANS has two principal branches.
   D) The postganglionic neuron lies entirely in the peripheral nervous system.
   E) All of the above are correct statements.
   Answer: B

8. Sympathetic fibers
   A) are unmyelinated fibers.
   B) originate in the brain.
   C) originate in the spinal cord.
   D) tend to be long at the preganglionic site.
   E) synapse in the terminal ganglion.
   Answer: C

9. Which of the following organs receives sympathetic innervation only?
   A) small intestines
   B) heart
   C) urinary bladder
   D) adrenal medulla
   E) sigmoid colon
   Answer: D

10. All of the following are involved in the control of the autonomic nervous system EXCEPT
    A) the cerebral cortex.
    B) the cerebellum.
    C) the medulla oblongata.
    D) thalamus.
    E) the hypothalamus.
    Answer: B

11. Which of the following are sympathetic ganglia?
    A) superior cervical ganglion
    B) prevertebral ganglia
    C) inferior mesenteric ganglion
    D) celiac ganglion
    E) all of the above
    Answer: E

12. All of the following are visceral effectors EXCEPT
    A) cardiac muscle.
    B) smooth muscle.
    C) skeletal muscle.
    D) glandular epithelium.
    E) all of the above are visceral effectors.
    Answer: C

13. In the autonomic nervous system there are ___ neurons between the CNS and the visceral effector.
    A) 5
    B) 4
    C) 3
    D) 2
    E) 1
    Answer: D

14. In the ANS how many ganglia are present between the CNS and the visceral effector?
    A) 1
    B) 2
    C) 3
    D) 4
    E) none
    Answer: A

15. The preganglionic fibers of the autonomic nervous system release
    A) noradrenaline.
    B) norephinephrine.
    C) acethylcholine.
    D) GABA.
    E) adrenaline.
    Answer: C

16. Preganglionic fibers of the parasympathetic division of the ANS synapse with the
    A) sympathetic trunk ganglia.
    B) celiac ganglion.
    C) mesenteric ganglion.
    D) terminal ganglia.
    E) prevertebral ganglia.
    Answer: D

17. All of the following are true about preganglionic fibers of the sympathetic division EXCEPT
    A) they are myelinated.
    B) they leave the spinal cord through the anterior root.
    C) they synapse with many postganglionic cell bodies.
    D) they are longer than the postganglionic fibers.
    E) they use acetylcholine as the neurotransmitter.
    Answer: D

18. The junction of an autonomic fiber and its effector is called
    A) neuromuscular junction.
    B) neuroglandular junction.
    C) neuroeffector junction.
    D) All of the above.
    E) None of the above.
    Answer: D

19. Adrenergic neurons are neurons that release
    A) acetylcholine.
    B) norepinephrine.
    C) histamine.
    D) Both A and B.
    E) All of the above.
    Answer: B

20. The neurotransmitter of postganglionic fibers of the sympathetic division is
    A) norepinephrine.
    B) acetylcholine.
    C) acetylcholinesterase.
    D) dopamine.
    E) GABA.
    Answer: A

21. The neurotransmitter of postganglionic fibers in the parasympathetic division is
    A) glycine.
    B) acetylcholine.
    C) acetylcholinesterase.
    D) dopamine.
    E) norepinephrine.
    Answer: B

22. Impulses for one division of the ANS stimulates an organ's activity while the other division inhibits the organ's activity. This is called:
    A) double innervation.
    B) dual innervation.
    C) autonomic innervation.
    D) synergistic innervation.
    E) agonistic innervation.
    Answer: B

23. The sympathetic division of the ANS
    A) increases the activities of the heart.
    B) decreases the activities of the heart.
    C) increases digestive processes.
    D) Both A and C.
    E) None of the above.
    Answer: A

24. All of the following are fight-or –flight responses EXCEPT
    A) increase of the heart rate.
    B) increase in blood pressure.
    C) constriction of the pupils.
    D) dilation of the bronchioles.
    E) decrease of digestive secretions.
    Answer:  C

25. Somatic motor neurons
    A) have long preganglionic fibers.
    B) release norepinephrine.
    C) release acetylcholine.
    D) innervate smooth muscle.
    E) are regulated by the hypothalamus.
    Answer:  C

26. The autonomic nervous system
    A) is under voluntary control.
    B) is always excitatory.
    C) has one motor neuron from the CNS to the effector.
    D) operates without conscious control.
    E) innervates skeletal muscles.
    Answer:  D

27. The portion of the nervous system regulating smooth muscle activity is the
    A) central nervous system.
    B) sympathetic nervous system.
    C) autonomic nervous system.
    D) somatic nervous system.
    E) somatosensory system.
    Answer:  E

28. The balance between sympathetic and parasympathetic activity is regulated by the
    A) hypothalamus.
    B) hippocampus.
    C) cerebellum.
    D) medulla.
    E) pituitary.
    Answer: A

29. Which of the following systems is responsible for skeletal muscle activity
    A) autonomic nervous system.
    B) sympathetic nervous system.
    C) parasympathetic nervous system.
    D) somatosensory system.
    E) somatic nervous system.
       Answer:  E

30. Autonomic fibers are collectively called
    A) somatic fibers.
    B) visceral motor fibers.
    C) visceral afferent fibers.
    D) visceral sensory fibers.
    E) cholinergic fibers.
    Answer: B

31. The divisions of the autonomic nervous system are called
    A) sympathetic and cholinergic divisions.
    B) parasympathetic and adrenergic divisions.
    C) cholinergic and adrenergic divisions.
    D) sympathetic and parasympathetic divisions.
    E) somatic and parasympathetic divisions.
    Answer: D

32. Which of the following are the neurotransmitters of the autonomic nervous system
    A) acetylcholine and dopamine.
    B) acetylcholine and norepinephrine.
    C) dopamine and epinephrine.
    D) glycine and GABA.
    E) norepinephrine only.
    Answer: B

33. "Dual innervation" means that glands and organs are innervated by both branches of the autonomic nervous system
    A) at the same time.
    B) with additive effects.
    C) but at opposing times for good balance.
    D) but the sympathetic dominates.
    E) but the parasympathetic dominates.
    Answer: C

34. The somatic and autonomic nervous system make up the _____ nervous system.
    A) involuntary
    B) peripheral
    C) central
    D) skeletal
    E) visceral
    Answer: B

35. Which of the following response is a sympathetic response?
    A) Increased bile secretion
    B) Breakdown of triglycerides in adipose tissue.
    C) Vasodilation in the penis.
    D) Promotion of insulin secretion.
    E) All of the above are due to sympathetic stimulation.
    Answer: B

36. Which of the following division(s) is/are called the craniosacral division?
    A) brain and spinal cord
    B) somatic division
    C) parasympathetic and sympathetic division
    D) parasympathetic division
    E) sympathetic division
    Answer: D

37. The ganglia which lie on either side of the backbone are the
    A) sympathetic trunk ganglia.
    B) terminal ganglia.
    C) dorsal root ganglia.
    D) prevertebral ganglia.
    E) superior mesenteric ganglia.
    Answer: A

38. All of the following are effects of the parasympathetic division EXCEPT
    A) conserving body energy.
    B) restoring of body energy.
    C) dilation of pupils.
    D) promoting digestion.
    E) increase in bile secretion.
    Answer: C

39. Widening of the coronary blood vessels is the result of _____ activity.
    A) somatic
    B) sympathetic
    C) parasympathetic
    D) somatosensory
    E) ganglionic
    Answer: B

40. Insulin secretion is promoted by the activity of the
    A) sympathetic nervous system.
    B) hypothalamus.
    C) somatic nervous system.
    D) parasympathetic nervous system.
    E) autonomic ganglia.
    Answer: D

41. The adrenal glands produce epinephrine and norepinephrine to intensify and prolong ___ effects.
    A) CNS
    B) endocrine
    C) somatic
    D) parasympathetic
    E) sympathetic
    Answer: E

42. Acetylcholine is inactivated by
    A)  pseudocholinesterase.
    B)  acetylcholinesterase.
    C)  acetase.
    D)  norepinephrine.
    E)  epinephrine.
    Answer:  B

43. The acronym "SLUDD" deals with
    A)  paradoxical fear.
    B)  parasympathetic tone.
    C)  "fight or flight" response.
    D)  Both A and B.
    E)  All of the above.
    Answer:  D

44. Massive activation of the parasympathetic division occurs in
    A)  the "fight or flight" response.
    B)  paradoxical fear.
    C)  exercise.
    D)  Both A and B.
    E)  All of the above.
    Answer:  B

45.  Excessive contraction of arterioles within the fingers and toes due to prolonged sympathetic stimulation can be
    A)  autonomic dysreflexia.
    B)  Horner's syndrome.
    C)  Downs syndrome.
    D)  Raynaud's disease.
    E)  Reyes syndrome.
    Answer:  D


## True-False

*Write T if the statement is true and F if the statement is false.*

1. The hypothalamus receives input from areas of the nervous system concerned with emotions.
   Answer:  True

2. The autonomic nervous system innervates the skeletal muscles.
   Answer:  False

3. The axon of a postganglionic neuron is called a postganglionic fiber.
   Answer:  True

4. Postganglionic fibers are myelinated and end in visceral effectors.
   Answer:  False

5.  The parasympathetic division of the ANS originates in the brain and spinal cord.
    Answer: True

6.  The sympathetic division is involved with energy expenditure.
    Answer: True

7.  Fear stimulates the parasympathetic division of the ANS.
    Answer: False

8.  The dilation of pupils is an autonomic response.
    Answer: True

9.  All preganglionic fibers of the ANS are cholinergic.
    Answer: True

10. The neurotransmitter of the parasympathetic postganglionic fibers release norepinephrine.
    Answer: False

11. Terminal ganglia are sympathetic ganglia.
    Answer: False

12. Parasympathetic ganglia are near or within visceral effectors.
    Answer: True

13. Parasympathetic stimulation increases sweat secretion.
    Answer: False

14. The parasympathetic branch of the autonomic nervous system prepares the body or the "fight-or-flight" response.
    Answer: False

15. Pre- and postganglionic fibers are present in the somatic nervous system.
    Answer: False

## Short Answer

*Write the word or phrase that best completes each statement or answers the question.*

1.  The subdivision of the autonomic nervous system are the ____ and ____ divisions.
    Answer: sympathetic and parasympathetic

2.  Autonomic fibers are also called ____.
    Answer: visceral motor fibers

3.  The sympathetic division is also called ____ division.
    Answer: thoracolumbar

4.  The celiac ganglion is a ____ ganglion.
    Answer: sympathetic

5. The part of the nervous system that regulates smooth and cardiac muscle is the _____ nervous system.
   Answer: autonomic

6. Cholinergic fibers release ____.
   Answer: acetylcholine

7. The enzyme that inactivates acetylcholine is ____.
   Answer: acetylcholinesterase

8. Synapses between autonomic neurons and their effectors are called _____ junctions.
   Answer: neuroeffector

9. The dilation of bronchioles is due to ____ activity.
   Answer: sympathetic

10. The somatic nervous system is under ____ control.
    Answer: voluntary

11. The neurotransmitter of all preganglionic fibers is ____.
    Answer: acetylcholine

12. The neurotransmitter of sympathetic postganglionic fibers is ____.
    Answer: norepinephrine

13. The ____ division of the ANS stimulates epinephrine and norepinephrine secretion.
    Answer: sympathetic

14. The ____ division of the ANS inhibits digestion.
    Answer: sympathetic

15. All the axons of the ANS are ____ fibers.
    Answer: motor

16. Organs that receive impulses from both divisions of the ANS have ____.
    Answer: dual innervation

17. Neurons that release acetylcholine are referred to as _____ neurons.
    Answer: cholinergic

18. Synapses of visceral motor fibers occur in ____.
    Answer autonomic ganglia

19. Long preganglionic fibers belong to the ____ division.
    Answer: parasympathetic

20. The neuroeffector junctions at glands are referred to as ____.
    Answer: neuroglandular junctions

21. Parasympathetic impulses ____ the rate of the heartbeat.
    Answer: decrease

22. During extreme stress the _____ division dominates the _____ division.
    Answer: sympathetic, parasympathetic

23. The division that conserves and restores the body's energy is the _____ division.
    Answer: parasympathetic

24. The acronym helpful in remembering five responses that occur when parasympathetic tone rises is
    "_____".
    Answer: "SLUDD"

25. Renin secretion in the kidney is a _____ response.
    Answer: sympathetic


## Matching

*Choose the item from Column 2 that best matches each item in Column 1.*

1. Column 1: Decreases heart rate
   Column 2: Parasympathetic

2. Column 1: Increases digestion
   Column 2: Parasympathetic

3. Column 1: Stimulates renin secretion
   Column 2: Sympathetic

4. Column 1: Dilation of pupils
   Column 2: Sympathetic

5. Column 1: Stimulates salivary gland secretion
   Column 2: Parasympathetic

6. Column 1: Inhibits gastric gland secretion
   Column 2: Sympathetic

7. Column 1: Increased respiratory rate
   Column 2: Sympathetic

8. Column 1: Erection of hairs
   Column 2: Sympathetic

9. Column 1: Promotes gastric gland secretion
   Column 2: Parasympathetic

10. Column 1: Constricts pupils
    Column 2: Parasympathetic

# Essay

*Write your answer in the space provided or on a separate sheet of paper.*

1. Name the divisions of the autonomic nervous system and compare their functions.
   Answer:  The parasympathetic division of the ANS controls bodily functions during normal activities, while the sympathetic nervous system takes over during stress and prepares the body for the "fight-or-flight" response.

2. Compare the length of the pre- and postganglionic fibers of the ANS divisions and name the neurotransmitters used in each of the fibers.
   Answer: Parasympathetic division:  long preganglionic fibers which use acetylcholine as neurotransmitter, and short postganglionic fibers which also use acetylcholine. Sympathetic division:  short preganglionic fibers using acetylcholine, and long post ganglionic fibers utilizing noradrenaline.

3. Compare the functional differences between the autonomic and somatic nervous system.
   Answer:  The somatic nervous system innervates skeletal muscles and is under voluntary control, the autonomic nervous system innervates cardiac muscle, smooth muscle, and glands.  The ANS is an involuntary system.

## CHAPTER 12     Sensations

## Multiple-Choice

*Choose the one alternative that best completes the statement or answers the question.*

1. An adequate stimulus applied to a sensory receptor will cause a(n)
   A) receptor potential.
   B) generator potential.
   C) synaptic potential.
   D) equilibrium potential.
   E) action potential.
   Answer:  B

2. The simplest sensory receptors are
   A) thermoreceptors.
   B) encapsulated nerve endings.
   C) proprioceptors.
   D) free nerve endings.
   E) touch receptors.
   Answer: D

3. A decreased sensitivity that occurs when a stimulus is continuously applied to sensory receptors
   is called
   A) adaptation.
   B) relaxation.
   C) accommodation.
   D) refraction.
   E) consolidation.
   Answer:  A

4. Which of the following senses does NOT belong to the group of somatic senses?
   A) touch receptors
   B) temperature receptors
   C) visceral receptors
   D) pressure receptors
   E) proprioceptors
   Answer: C

5. All of the following are cutaneous sensations EXCEPT
   A) touch.
   B) pressure.
   C) temperature.
   D) equilibrium.
   E) vibration.
   Answer:  D

6.  All of the following are mechanoreceptors EXCEPT
    A) hair root plexuses.
    B) Meissner's corpuscles.
    C) nociceptors.
    D) Pacinian corpuscles.
    E) organs of Ruffini.
    Answer:  C

7.  Which of the following senses falls under the category of complex receptors?
    A) taste
    B) touch
    C) heat
    D) cold
    E) pain
    Answer: A

8.  Touch receptors are classified as
    A) complex receptors.
    B) nociceptors.
    C) interoceptors.
    D) exteroceptors.
    E) proprioceptors.
    Answer: D

9.  Pressure receptors, which are located around joints, tendons, and muscles, are
    A) Pacinian corpuscles.
    B) Meissner's corpuscle.
    C) organs of Ruffini.
    D) tactile receptors.
    E) nociceptors.
    Answer:  A

10. The sensation of *itch* results from the stimulation of
    A) Meissner's corpuscles.
    B) Pacinian corpuscles.
    C) free nerve endings.
    D) proprioceptors.
    E) tactile receptors.
    Answer: C

11. Nociceptors are stimulated by
    A) acetylcholine.
    B) histamine.
    C) prostaglandins.
    D) renin.
    E) norepinephrine.
    Answer: C

12. A kind of pain often experienced by people with an amputated limb is called
    A) referred pain.
    B) delayed pain.
    C) dull pain.
    D) sharp pain.
    E) phantom pain.
    Answer: E

13. The sense of muscle activity is called
    A) proprioception.
    B) nociception.
    C) stretch reception.
    D) adaptation.
    E) gustation.
    Answer: A

14. All of the following are proprioceptors EXCEPT
    A) muscle spindles.
    B) joint kinesthetic receptors.
    C) Golgi tendon organs.
    D) Ruffini organs.
    E) maculae and cristae.
    Answer: D

15. Muscle spindles are examples of
    A) free nerve endings.
    B) proprioceptors.
    C) Pacinian corpuscles.
    D) itch receptors.
    E) None of the above.
    Answer: B

16. Olfactory receptors
    A) are located in the nasal cavity.
    B) are stimulated by gaseous particles.
    C) adapt.
    D) have knob-shaped dendrites.
    E) all of the above are correct.
    Answer: E

17. Which of the following areas of the brain are involved in olfaction?
    A) temporal lobe
    B) hypothalamus
    C) limbic system
    D) none of the above
    E) all of the above
    Answer: E

18. Taste receptors are called
    A) taste buds.
    B) gustatory receptors.
    C) olfactory receptors.
    D) filiform papillae.
    E) taste pores.
    Answer: B

19. The region that is most sensitive to bitter taste is the
    A) tip of the tongue.
    B) back of the tongue.
    C) middle of the tongue.
    D) margin of the tongue.
    E) side of the tongue.
    Answer: B

20. The cranial nerves involved in the gustatory pathway are the
    A) glossopharyngeal, vagus, and gustatory.
    B) optic, glossopharyngeal, and facial.
    C) vagus, hypoglossal, and abducens.
    D) facial, glossopharyngeal, and vagus.
    E) vestibulochochlear, abducens, and hypoglossal.
    Answer: D

21. A physician specialized in the diagnosis and treatment of eye disorders with drugs, surgery, and corrective lenses is a(n)
    A) oncologist.
    B) pharmacologist.
    C) optician.
    D) ophthalmologist.
    E) optometrist.
    Answer: D

22. The immediate structure leading tears away from the lacrimal glands is/are the
    A) nasolacrimal duct.
    B) lacrimal ducts.
    C) eustachian tube.
    D) nasal cavity.
    E) nasal concha.
    Answer: B

23. The bactericidal enzymes present in tears are
    A) lysosomes.
    B) lysozyme.
    C) bacteriozyme.
    D) lipase.
    E) protease.
    Answer: B

24. The "white" of the eye is called
    A) conjunctiva.
    B) sclera.
    C) retina.
    D) cornea.
    E) iris.
    Answer: B

25. Which of the following belongs to the vascular tunic of the eye
    A) cornea.
    B) sclera.
    C) retina.
    D) iris.
    E) pupil.
    Answer: D

26. The muscle responsible for the change of the shape of the lens is the
    A) arrector pili muscle.
    B) orbicularis oculi.
    C) ciliary muscle.
    D) superior rectus muscle.
    E) retinal muscle.
    Answer: C

27. The layer that contains photoreceptors is the
    A) iris.
    B) retina.
    C) sclera.
    D) cornea.
    E) iris.
    Answer: B

28. The area containing the highest concentration of cones is the
    A) central fovea.
    B) optic disc.
    C) iris.
    D) macula lutea.
    E) pupil.
    Answer: A

29. The clear jellylike substance behind the lens of the eye is the
    A) aqueous humor.
    B) cellular body.
    C) ciliary body.
    D) anterior humor.
    E) vitreous humor.
    Answer: E

30. The bending of light that happens at the cornea and lens is
    A) accommodation.
    B) reflection.
    C) contraction.
    D) refraction.
    E) adaptation.
    Answer: D

31. The inability to clearly see near objects is called
    A) nearsightedness.
    B) astigmatism.
    C) hypermetropia.
    D) presbyopia.
    E) myopia.
    Answer: C

32. The "canal of Schlemm" is another term for the
    A) vitreous chamber.
    B) vitreous body.
    C) anterior cavity.
    D) aqueous humor.
    E) scleral venous sinus.
    Answer: E

33. The photopigment in rods is
    A) chlorophyll.
    B) rhodopsin.
    C) opsin.
    D) melanin.
    E) retinal.
    Answer: B

34. Which of the following form the optic nerve
    A) ganglion cells.
    B) retinal cells.
    C) bipolar cells.
    D) rods.
    E) cones.
    Answer: A

35. All of the following belong to the external ear EXCEPT
    A) auditory tube.
    B) auricle.
    C) external auditory canal.
    D) tympanic membrane.
    E) ceruminous glands.
    Answer: A

36. Which of the following belongs to the middle ear?
    A) cochlea.
    B) ossicles.
    C) bony labyrinth.
    D) vestibule.
    E) scala tympani.
    Answer: A

37. The middle part of the bony labyrinth is the
    A) cochlea.
    B) saccule.
    C) utricle.
    D) vestibule.
    E) ampulla.
    Answer: D

38. Located between the cochlear duct and the scala vestibuli is the
    A) vestibular membrane.
    B) tectorial membrane.
    C) basement membrane.
    D) cochlear duct.
    E) basilar membrane.
    Answer: A

39. The intensity of sound is measured in
    A) herz.
    B) decibel.
    C) frequencies.
    D) Both A and C.
    E) Both B and C
    Answers: B

40. The center of the eardrum is connected to the
    A) incus.
    B) cochlea.
    C) vestibule.
    D) malleus.
    E) staples.
    Answer: D

41. The receptors for hearing are located in the
    A) semicircular canals.
    B) cochlea.
    C) utricle.
    D) sacculae.
    E) vestibule.
    Answer: B

42. The senses for dynamic equilibrium are located in the
    A) utricle.
    B) sacculae.
    C) Organ of Corti.
    D) cochlea.
    E) semicircular ducts.
   Answer: E

43. An abnormally high intraocular pressure is referred to as
    A) cataract.
    B) glaucoma.
    C) trachoma.
    D) vertigo.
    E) nystagmus.
   Answer: B

44. The most common cause of blindness in the United States is
    A) scotoma.
    B) trachoma.
    C) cataracts.
    D) glaucoma.
    E) macular degeneration.
   Answer: D

45. "Pinkeye" is the common term for
    A) trachoma.
    B) conjunctivitis.
    C) myringitis.
    D) blepharitis.
    E) keratitis.
   Answer: B


## True-False

*Write T if the statement is true and F if the statement is false.*

1. The distinct quality that makes one sensation different from others is its modality.
   Answer: True

2. Muscle spindles are found at the junction of a tendon with a muscle.
   Answer: False

3. Equilibrium is classified as a special sense.
   Answer: True

4. Cutaneous receptors consist of the axons of sensory neurons.
   Answer: False

5. Crude touch refers to the feeling that something has touched the skin and the exact location can be determined.
   Answer: False

6. Thermoreceptors are free nerve endings.
   Answer: True

7. Pain is essential because it protects us from greater damage.
   Answer: True

8. The highest concentration of taste buds is found in filiform papillae.
   Answer: False

9. The adult eyeball measures about 3.5 cm in diameter.
   Answer: False

10. The middle layer of the eyeball is the vascular tunic.
    Answer: True

11. Cones are specialized for color vision.
    Answer: True

12. Night blindness is most often caused by a vitamin C deficiency.
    Answer: False

13. Some fibers of the optic nerve cross in the optic tract.
    Answer: False

14. The middle ear is a small air-filled cavity between the eardrum and the internal ear.
    Answer: True

15. The walls of the utricle contain a small, flat region called macula.
    Answer: True

## Short Answer

*Write the word or phrase that best completes each statement or answers the question.*

1. The depolarization of the membrane in sensory receptors is called _____ _____.
   Answer: generator potential

2. The small area of the retina, which does not contain photoreceptors but the optic nerve, is the ____.
   Answer: optic disc

3. Receptors that detect pressure or stretching are collectively called _____.
   Answer: mechanoreceptors

4. The conscious or subconscious awareness of external or internal conditions of the body is a(n) _____.
   Answer: sensation

5. Conscious sensations are integrated in the _____ _____.
   Answer: cerebral cortex

6. The ability to recognize which point of the body is touched is referred to as ____.
   Answer:  discriminative touch

7. Receptors for pain are called _____.
   Answer: nociceptors

8. Receptors for taste are called _____ receptors.
   Answers: gustatory

9. The ___ gland produces tears.
   Answer: lacrimal

10. The colored portion of the eyeball is the _____.
    Answer: iris

11. A loss of transparency of the lens is known as _____.
    Answer: cataract

12. The ability of the lens to change its curvature is called _____.
    Answer: accommodation

13. Images that focus upside down on the retina are called _____ images.
    Answer: inverted

14. The receptors for color vision and bright light are _____.
    Answer: cones

15. _____ glands produce earwax.
    Answer: ceruminous

16. The bony labyrinth contains fluid called the _____.
    Answer: perilymph

17. The organ of hearing is the _____.
    Answer: Organ of Corti (spiral organ)

18. The balance and posture of the body without movement of the head is called _____.
    Answer: static equilibrium

19. An inflammation of the conjunctiva is called _____.
    Answer: conjunctivitis

20. The opening of a taste bud is the _____.
    Answer: taste pore

21. The hole in the center of the iris is the _____.
    Answer: pupil

22. The pressure in the eye that is mainly produced by the aqueous humor is the _____.
    Answer: intraocular pressure

23. The thin membrane between the external auditory canal and the ossicles is the _____.
    Answer: tympanic membrane

24. The greater the frequency of sound the higher the _____.
    Answer: pitch

25. A type of receptor that provides information about the internal environment is called a(n) _____.
    Answer: interoceptor


## Matching

*Choose the item from Column 2 that best matches each item in Column 1.*

1. Column 1: An acute bacterial infection of the middle ear.
   Column 2: otitis media

2. Column 1: A disorder brought on by motion.
   Column 2: motion sickness

3. Column 1: An inflammation of the eyelid.
   Column 2: blepharitis

4. Column 1: An inflammation of the auditory tube.
   Column 2: eustachitis

5. Column 1: An inflammation of the inner ear.
   Column 2: labyrinthitis

6. Column 1: A rapid involuntary movement of the eyeballs.
   Column 2: nystagmus

7. Column 1: Dilated pupil.
   Column 2: mydriasis

8. Column 1: A ringing in the ears.
   Column 2: tinnitus

9. Column 1: A change in the environment capable of activating sensory neurons.
   Column 2: stimulus

10. Column 1: The receptors of pain.
    Column 2: nociceptors

11. Column 1: A pain in the skin overlying the stimulated organ.
    Column 2: referred pain

12. Column 1: The lining of the eyelid.
    Column 2: conjunctiva

13. Column 1: Responsible for night vision.
    Column 2: Rods

14. Column 1: Responsible for dynamic equilibrium.
    Column 2: semicircular ducts

15. Column 1: Corpuscles of touch.
    Column 2: Meissner's corpuscles


## Essay

*Write your answer in the space provided or on a separate sheet of paper.*

1. Define the term sensation.
   Answer: Sensation refers to the awareness of external or internal conditions of the body.

2. Name the proprioceptive receptors and briefly describe the location of each.
   Answer: Muscle spindles: located between skeletal muscle fibers. Tendon organs: located at the junction of a tendon with muscle. Join kinesthetic receptors: located around synovial joints. Maculae and Cristae: located in the inner ear.

3. Describe the olfactory pathway.
   Answer: Axons of the olfactory receptors form the olfactory nerves, the olfactory nerves convey impulses to the olfactory bulbs, and then to the olfactory tract, which sends the impulses to t the primary olfactory area in the temporal lobe of the cerebral cortex.

4. Name the three coats of the eye and structures that belong to each coat.
   Answer: Fibrous tunic: sclera and cornea. Vascular tunic: choroids, ciliary body, and iris. Retina: photoreceptor cells (rods and cones), bipolar cells, ganglion cells.

# CHAPTER 13    The Endocrine System

## Multiple-Choice

*Choose the one alternative that best completes the statement or answers the question.*

1.  Which of the following systems work closely with the endocrine system to coordinate the body's functions?
    A)  Cardiovascular system
    B)  Respiratory system
    C)  Reproductive system
    D)  Nervous system
    E)  Lymphatic system
    Answer: D

2.  All of the following are endocrine glands EXCEPT:
    A)  adrenal glands.
    B)  sebaceous glands.
    C)  parathyroid glands.
    D)  pineal glands.
    E)  pituitary glands.
    Answer: B

3.  Which of the following organs contain hormone producing cells?
    A)  stomach
    B)  liver
    C)  skin
    D)  All of the above.
    E)  None of the above.
    Answer: D

4.  A second messenger in the action of many water-soluble hormones such as peptides is
    A)  tRNA.
    B)  ATP.
    C)  cAMP.
    D)  PTH.
    E)  CoA.
    Answer:  C

5.  Prostaglandins chemically belong to the class of
    A)  steroid hormones.
    B)  eicosanoids.
    C)  peptides.
    D)  proteins.
    E)  amines.
    Answer: B

6. Which of the following hormones is a lipid soluble substance?
   A) estrogen
   B) melatonin
   C) histamine
   D) calcitonin
   E) gastrin
   Answer: A

7. Hormones that bind to receptors within the target cells are
   A) lipid soluble.
   B) water soluble.
   C) second messengers.
   D) proteins.
   E) prohormones.
   Answer: A

8. The hormones regulating blood calcium levels are
   A) insulin and glucagon.
   B) glycogen and PTH.
   C) inhibiting hormones.
   D) PTH and calcitonin.
   E) calcitonin and ATP.
   Answer: D

9. A hormone that under certain circumstances is regulated by positive feedback is
   A) calcitonin.
   B) histamine.
   C) oxytocin.
   D) melatonin.
   E) insulin.
   Answer: C

10. The pituitary gland is attached to the hypothalamus by the
    A) epithalamus.
    B) infundibulum.
    C) parafollicular cells.
    D) intermediate mass.
    E) corpus callosum.
    Answer: B

11. Which of the following hormones stimulates testosterone production by the testis?
    A) TSH
    B) FSH
    C) ACTH
    D) LH
    E) GH
    Answer: D

12. Which of the following hormones is released in response to a nerve impulse?
    A) epinephrine
    B) cortisol
    C) testosterone
    D) insulin
    E) glucagon
    Answer: A

13. All of the following are hormones of the anterior pituitary EXCEPT
    A) human growth hormone.
    B) follicle-stimulating hormone.
    C) adrenocorticotropic hormone.
    D) prolactin.
    E) oxytocin.
    Answer: E

14. The hormone which effects skin pigmentation is
    A) MSH.
    B) FSH.
    C) GH.
    D) LH.
    E) ACTH.
    Answer: A

15. Which of the following hormones is released in response to hypoglycemia
    A) TSH.
    B) FSH.
    C) LH.
    D) GH.
    E) MSH.
    Answer: D

16. Insulinlike growth factor is released in response to
    A) insulin.
    B) glucagon.
    C) human growth hormone.
    D) luteneizing hormone.
    E) all pituitary hormones.
    Answer: C

17. All of the following hormones are released in response to releasing hormones EXCEPT
    A) thyroid stimulating hormone.
    B) human growth hormone.
    C) follicle stimulating hormone.
    D) prolactin.
    E) oxytocin.
    Answer: E

18. Which of the following hormones do neurosecretory cells produce?
    A) antidiuretic hormone
    B) calcitonin
    C) insulin
    D) growth hormone
    E) adrenocorticotropic hormone
    Answer: A

19. Antidiuretic hormone and oxytocin are produced by the
    A) posterior pituitary.
    B) anterior pituitary.
    C) thyroid.
    D) hypothalamus.
    E) pancreas.
    Answer: D

20. The hormone which is released in large quantities just prior to childbirth is
    A) estrogen.
    B) oxytocin.
    C) prolactin.
    D) progesterone.
    E) growth hormone.
    Answer: B

21. Which of the following hormones controls the production and release of glucocoticoids?
    A) ADH
    B) ACTH
    C) GH
    D) FSH
    E) LH
    Answer: B

22. The lobes of the thyroid gland are connected by a tissue mass called the
    A) body.
    B) infundibulum.
    C) cortex.
    D) stalk.
    E) isthmus.
    Answer: E

23. Calcitonin is a hormone of the
    A) adrenal cortex.
    B) thyroid gland.
    C) parathyroid gland.
    D) pituitary gland.
    E) thymus gland.
    Answer: B

24. Which of the following cells produce thyroxine?
    A) alpha cells
    B) oxyphil cells
    C) neurosecretory cells
    D) follicular cells
    E) parafollicular cells
    Answer: D

25. The hormone released in response to low blood calcium levels is
    A) oxytocin.
    B) parathyroid hormone.
    C) thyroxine.
    D) growth hormone.
    E) calcitonin.
    Answer: B

26. The hormone that inhibits the action of osteoclasts is
    A) oxytocin.
    B) vasopressin.
    C) calcitonin.
    D) parathyroid hormone.
    E) thyroxine.
    Answer: C

27. Mineralcorticoids
    A) are produced in the adrenal cortex.
    B) are steroid hormones.
    C) help regulate the homeostasis of sodium and potassium.
    D) All of the above.
    E) None of the above.
    Answer: D

28. Glucocorticoids are steroid hormones produced by the
    A) ovaries.
    B) testis.
    C) adrenal gland.
    D) thyroid gland.
    E) hypothalamus.
    Answer: C

29. Which of the following endocrine glands is directly controlled by the autonomic nervous system?
    A) adrenal medulla
    B) adrenal cortex
    C) anterior pituitary
    D) thyroid gland
    E) thymus gland
    Answer: A

30. Which of the following hormones are responsible for the fight-or-flight response?
    A) epinephrine
    B) norepinephrine
    C) acetylcholine
    D) A and B
    E) B and C
    Answer: D

31. The gland which can be classified as an endocrine and an exocrine glands is the
    A) thyroid.
    B) thymus.
    C) pancreas.
    D) pituitary.
    E) hypothalamus.
    Answer: C

32. Insulin is secreted by
    A) alpha cells.
    B) beta cells.
    C) delta cells.
    D) F-cell.
    E) chief cells.
    Answer: B

33. Glucagon
    A) accelerates the conversion of glycogen into glucose.
    B) slows down glucose formation from lactic acid.
    C) decreases the conversion of glycogen into glucose.
    D) speeds up protein synthesis within cells.
    E) accelerates the transport of glucose from blood into cells.
    Answer: A

34. The development and maintenance of the female sex characteristics is the responsibility of
    A) estrogen and androgen.
    B) progesterone and testosterone.
    C) relaxin and inhibin.
    D) progesterone and relaxin.
    E) progesterone and estrogen.
    Answer: E

35. FSH secretion is inhibited by
    A) relaxin.
    B) testosterone.
    C) LH.
    D) inhibin.
    E) androgen.
    Answer: D

36. Melatonin is a hormone of the
    A) pituitary gland.
    B) pancreas.
    C) pineal gland.
    D) thymus gland.
    E) adrenal gland.
    Answer: C

37. Prostaglandins
    A) act as local hormones.
    B) help induce inflammation.
    C) are important in fat metabolism.
    D) All of the above.
    E) None of the above.
    Answer: D

38. The second stage of GAS (general adaptation syndrome) is
    A) the stressor.
    B) the resistance reaction.
    C) the alarm reaction.
    D) the flight-or-fight response.
    E) the stage of exhaustion.
    Answer: B

39. Oversecretion of hGH during childhood is called
    A) giantism.
    B) acromegaly.
    C) pituitary dwarfism.
    D) cretinism.
    E) goiter.
    Answer: B

40. Insufficient ADH release causes
    A) diabetes insipidus.
    B) diabetes mellitus.
    C) tetany.
    D) cretinism.
    E) aldosteronism.
    Answer: A

41. Cushing's syndrome is due to oversecretion of
    A) glucocorticoids.
    B) testosterone.
    C) ADH.
    D) mineralcorticoids.
    E) thyroxine.
    Answer: A

42. A decline in the number of pancreatic beta cells is characteristic for
    A) diabetes insipidus.
    B) type II diabetes.
    C) type I diabetes.
    D) maturity onset diabetes.
    E) noninsulin-dependent diabetes.
    Answer: C

43. Cells that respond to a particular hormone are called
    A) receptor cells.
    B) effector cells.
    C) secretory cells.
    D) target cells.
    E) active cells.
    Answer: D

44. Hormone secretion is controlled by all of the following EXCEPT
    A) nerve impulses.
    B) circulating chemicals.
    C) releasing hormones.
    D) inhibiting hormones.
    E) All of the above.
    Answer: E

45. The activities of the adrenal cortex are controlled by
    A) FSH.
    B) LH.
    C) hGH.
    D) MSH.
    E) ACTH.
    Answer: E

## True-False

*Write T if the statement is true and F if the statement is false.*

1. The nervous system and the endocrine system regulate homeostasis.
    Answer: True

2. Hormones do not regulate the activity of the immune system.
    Answer: False

3. Steroid hormones are derivatives of cholesterol.
    Answer: True

4. Most peptide and protein hormones diffuse through the phospholipid bilayer to get into a target cell.
    Answer: False

5. Hormone secretion is primarily regulated by negative feedback mechanisms.
    Answer: True

6. The hypothalamus has been nicknamed "master gland".
   Answer: False

7. Cells of the posterior pituitary produce vasopressin.
   Answer: False

8. Prolactin is a hormone involved in the initiation of milk production in the mammary glands.
   Answer: True

9. Hormones of the hypothalamus control the release of human growth hormone.
   Answer: True

10. FSH stimulates the testis to produce testosterone.
    Answer: False

11. Vasopressin is another name for antidiuretic hormone.
    Answer: True

12. Alcohol inhibits ADH secretion resulting in an increased urine output.
    Answer: True

13. Calcitonin is a hormone of the parathyroid gland.
    Answer: False

14. The renin-angiotensin pathway is involved in the control of aldosterone secretion.
    Answer: True

15. Androgens are hormones secreted by the male endocrine system only.
    Answer: False

## Short Answer

*Write the word or phrase that best completes each statement or answers the question.*

1. The science concerned with the endocrine system is called ____.
   Answer: endocrinology

2. The endocrine system releases messenger molecules called ____.
   Answer: hormones

3. Hormones that pass from secretory cells into the interstitial fluid and then into the blood are called _____ hormones.
   Answer: circulating

4. Chemically leukotrienes are classified as _____ hormones.
   Answer: eicosanoid

5. The hypothalamus controls the activity of the anterior pituitary by way of _____.
   Answer: releasing and inhibiting hormones

6. The production of oocytes by the ovaries is stimulated by _____.
   Answer: follicle stimulating hormone (FSH)

7. Calcitriol is a hormone of the _____.
   Answer: kidneys

8. Hormones that can enter the plasma membrane without a receptor are _____-_____ hormones.
   Answer: lipid-soluble

9. The production and secretion of thyroid hormone is stimulated by _____.
   Answer: thyroid stimulating hormone (TSH)

10. The formation of corpus luteum is stimulated by _____.
    Answer: luteinizing hormone (LH)

11. The release of milk by the mammary glands depends on the hormone _____.
    Answer: oxytocin

12. The receptors in the hypothalamus that detect low water concentration in the blood are _____.
    Answer: osmoreceptors

13. The inhibition of bone breakdown is the function of _____.
    Answer: calcitonin

14. Oxyphyl cells are found in the _____ gland.
    Answer: parathyroid

15. The class of adrenal cortex hormones that deals with metabolism and resistance to stress are _____.
    Answer: glucocorticoids

16. The two principal hormones of the adrenal medulla are _____ and _____.
    Answer: epinephrine and norepinephrine

17. Mineralcorticoids are produced in the _____ _____.
    Answer: adrenal medulla

18. The endocrine portion of the pancreas consists of clusters of cells called _____.
    Answer: Islets of Langerhans (pancreatic islets)

19. The conversion of glucose into glycogen is accelerated by _____.
    Answer: insulin

20. Oversecretion of hGH during childhood results in _____.
    Answer: giantism

21. An enlarged thyroid gland is called _____.
    Answer: goiter

22. Hormones that influence other endocrine glands are called _____ hormones.
    Answer: tropic

23. Oxytocin and antidiuretic hormone are produced in cells of the _____.
    Answer: hypothalamus

24. Thyroid hormones regulate growth and development, the activity of the nervous system, and ____.
    Answer: metabolism

25. The adrenal gland is composed of the adrenal medulla and the ____.
    Answer: adrenal cortex

## Matching

*Choose the item from Column 2 that best matches each item in Column 1.*

1.  Column 1: Help induce inflammation.
    Column 2: prostaglandins

2.  Column 1: Controls general body growth and metabolism.
    Column 2: hGH

3.  Column 1: Stimulates the adrenal cortex.
    Column 2: ACTH

4.  Column 1: Stimulates the development of the ovarian follicles.
    Column 2: FSH

5.  Column 1: Initiates and maintains milk production by the mammary glands.
    Column 2: prolactin

6.  Column 1: Decreases urine volume.
    Column 2: ADH

7.  Column 1: Contains iodine.
    Column 2: thyroid hormone

8.  Column 1: Are anti-inflammatory compounds.
    Column 2: glucocorticoids

9.  Column 1: Secreted by alpha cells.
    Column 2: glucagon

10. Column 1: Secreted by the pineal gland.
    Column 2: melatonin

**Essay**

*Write your answer in the space provided or on a separate sheet of paper.*

1. Name the different chemical classes of hormones and give three examples of each.
    Answer: 1.  Lipid derivatives: steroid hormones (aldosterone, cortisol, androgens, testosterone, estrogen, and progesterone); eicosanoids (prostaglandins, leukotrienes).
        2.  Amino acid derivatives: thyroid hormones epinephrine, norepinephrine, histamine, serotonin, and melatonin.
        3.  Peptides and proteins: releasing and inhibiting hormones, oxytocin, antidiuretic hormone, anterior pituitary hormones, insulin, glucagons, parathyroid hormone, calcitonin....

2. Name the seven broad areas of hormone effects.
    Answer: 1.  Regulation of the chemical composition and volume of internal environment.
        2.  Regulation of metabolism and energy balance.
        3.  Regulation of smooth muscle and    cardiac muscle contraction.
        4.  Maintaining of homeostasis.
        5.  Regulation of certain immune system activities.
        6.  Regulation of growth and development. 7. Regulation of reproductive activities.

3. Name the two basic mechanisms by which hormones can act on their target cells.
    Answer:  Lipid-soluble hormones enter the target cell and bind to a receptor within the cell.  The activated receptor will then alter cell function via the genes.  Water-soluble hormones will bind to a membrane receptor of the target cell and activate a second messenger that is cyclic AMP. This activation will eventually produce the physiological response of the target cell.

## CHAPTER 14    The Cardiovascular System: Blood

## Multiple-Choice

*Choose the one alternative that best completes the statement or answers the question.*

1.  Which of the following are functions of the blood?
    A)  transportation
    B)  regulation
    C)  protection
    D)  A and B
    E)  All of the above
    Answer: E

2.  The blood volume of an averaged sized male is
    A)  3 to 4 liters.
    B)  4 to 5 liters.
    C)  5 to 6 liters.
    D)  6 to 7 liters.
    E)  6 liters exactly.
    Answer: C

3.  The pH range for blood is
    A)  7.45 – 7.65.
    B)  7.35 – 7.45.
    C)  7.25 – 7.35.
    D)  7.15 – 7.45.
    E)  7.00 – 7.35.
    Answer:  B

4.  The thin layer of white blood cells in centrifuged blood is called
    A)  hematocrit.
    B)  plasma.
    C)  serum.
    D)  buffy coat.
    E)  matrix of blood.
    Answer: D

5.  Most of the plasma proteins are
    A)  albumins.
    B)  fibrinogens.
    C)  gamma globulins.
    D)  alpha globulins.
    E)  beta globulins.
    Answer: A

6.  Which of the following belongs to agranular leukocytes?
    A)  thrombocyte
    B)  neutrophil
    C)  basophil
    D)  platelet
    E)  monocyte
    Answer:  E

7. Which of the following cells do NOT have a nucleus?
   A) erythrocytes
   B) granulocytes
   C) leukocytes
   D) monocytes
   E) agranulocytes
   Answer: A

8. The life span of red blood cells is
   A) 150 days.
   B) 140 days.
   C) 130 days.
   D) 120 days.
   E) 100 days.
   Answer: D

9. The pigment in red blood cells that carries oxygen is
   A) erythropoietin.
   B) melatonin.
   C) hemoglobin.
   D) uribilonogen.
   E) biliverdin.
   Answer: C

10. Worn-out red blood cells are phagocytized in the
    A) liver.
    B) spleen.
    C) red bone marrow.
    D) All of the above.
    E) None of the above.
    Answer: D

11. Which of the following components of hemoglobin can be reused by other cells for protein synthesis?
    A) the heme group
    B) globin
    C) transferin
    D) iron
    E) stercobilin
    Answer: B

12. Cellular oxygen deficiency is called
    A) ischemia.
    B) hypoxia.
    C) jaundice.
    D) anemia.
    E) cyanosis.
    Answer: B

13. A test that measures the rate of erythropoiesis is called
    A) reticulocyte count.
    B) leukocyte count.
    C) differential count.
    D) hemoglobin count.
    E) hematocrit count.
    Answer: A

14. What is the substance some athletes use for blood "doping"?
    A) hemoglobin
    B) iron
    C) erythropoietin
    D) transferrin
    E) adrenaline
    Answer: C

15. Blood cells that contain blue-purple stained granules are
    A) thrombocytes.
    B) eosinophils.
    C) monocytes.
    D) neutrophils.
    E) basophils.
    Answer: E

16. To maintain normal quantities of erythrocytes, the body must produce _____ new mature cells per second.
    A) 500,000
    B) 700,000
    C) 1 million
    D) 2 million
    E) 3 million
    Answer: D

17. The first phagocytotic cell at the site of a bacterial invasion are
    A) lymphocytes.
    B) neutrophils.
    C) eosinophils.
    D) thrombocytes.
    E) monocytes.
    Answer: B

18. An allergic condition or a parasite infection is often manifested in a high count of
    A) eosinophils.
    B) neutrophils.
    C) basophils.
    D) monocytes.
    E) lymphocytes.
    Answer: A

19. Which of the following blood cells can develop into "wandering macrophages?
   A) neutrophils
   B) eosinophils
   C) basophils
   D) lymphocytes
   E) monocytes
   Answer: E

20. An increase in the number of white blood cells is called
   A) anemia.
   B) leukopenia.
   C) leukemia.
   D) leukocytosis.
   E) mononucleosis.
   Answer: D

21. Abnormally low levels of white blood cells which may be caused by radiation is called
   A) leukopenia.
   B) anemia.
   C) leukocytosis.
   D) leukemia.
   E) polycythemia.
   Answer: A

22. How many platelets should appear in each µL of blood?
   A) 20,000 – 40,000
   B) 100,000 – 150,000
   C) 200,000 – 250,000
   D) 150,000 – 400,000
   E) 400,000 – 600,000
   Answer: D

23. Which of the following blood proteins functions in blood-clotting?
   A) albumin
   B) fibrinogen
   C) alpha globulin
   D) beta globulin
   E) gamma globulin
   Answer: B

24. Which of the following sequences correctly describes the steps in the formation of a platelet plug?
   A) platelet release reaction, platelet aggregation, platelet adhesion
   B) platelet aggregation, platelet adhesion, platelet release reaction
   C) platelet adhesion, platelet release reaction, platelet aggregation
   D) platelet release reaction, platelet adhesion, platelet aggregation
   E) platelet adhesion, platelet aggregation, platelet release reaction
   Answer: C

25. The process of clotting in an unbroken blood vessel is called
    A) hemostasis.
    B) coagulation.
    C) thrombosis.
    D) vascular spasm.
    E) fibinolysis.
    Answer: B

26. Stoppage of bleeding is called
    A) hemostasis.
    B) vascular spasm.
    C) thrombosis.
    D) coagulation.
    E) embolism.
    Answer: A

27. The threads of a blood clot are formed by
    A) thrombin.
    B) prothrombin activator.
    C) platelet plug.
    D) fibrinogen.
    E) fibrin.
    Answer: E

28. All of the following are involved in blood clotting EXCEPT
    A) globulin.
    B) thrombin.
    C) calcium.
    D) prothrombinase.
    E) fibrin.
    Answer: A

29. Heredity deficiencies of coagulation is referred to as
    A) anemia.
    B) hemophilia.
    C) hemolysis.
    D) polycythemia.
    E) leukemia.
    Answer: B

30. The enzyme responsible for breaking up a blood clot is
    A) fibrinogen.
    B) ATPase.
    C) coagulase.
    D) plasminogen.
    E) plasmin.
    Answer: E

31. The anticoagulant produced by mast cells is
    A) plasminogen.
    B) plasmin.
    C) fibrinogen.
    D) heparin.
    E) vitamin K.
    Answer: D

32. A blood clot transported by the blood stream is a(n)
    A) platelet plug.
    B) thrombus.
    C) embolus.
    D) thrombin clot.
    E) None of the above.
    Answer: C

33. A person with blood type A has
    A) B antigens on the red blood cells.
    B) A antibodies in the plasma.
    C) A antigens on the red blood cells.
    D) Rh antigen on the red blood cells.
    E) A and B antibodies in the plasma.
    Answer: C

34. An individual with which ABO blood type can theoretically donate blood to recipients of all ABO types
    A) type A.
    B) type B.
    C) AB.
    D) type O.
    E) All of the above.
    Answer: D

35. The clumping of red blood cells due to incompatible blood transfusion is called
    A) hemolysis.
    B) embolism.
    C) agglutination.
    D) reaction.
    E) infiltration.
    Answer: C

36. Small inappropriate blood clots inside of a blood vessel are dissolved in a process called
    A) fibrinolysis.
    B) plasmolysis.
    C) clot retraction.
    D) thrombosis.
    E) agglutination.
    Answer: A

37. The most common blood type in Native Americans is
    A) A.
    B) B.
    C) AB.
    D) O.
    E) A an B.
    Answer: D

38. Insufficient production of erythrocytes due to lack of vitamin B12 is the cause for
    A) polycythemia.
    B) leukemia.
    C) aplastic anemia.
    D) pericious anemia.
    E) hemorrhagic anemia.
    Answer: D

39. The iron-containing portion of the hemoglobin molecule is
    A) ferrin.
    B) globin.
    C) heme.
    D) globulin.
    E) oxygen.
    Answer: C

40. The hormone released by the kidney and is stimulating RBC production is
    A) erythropoietin.
    B) plasmin.
    C) albumin.
    D) hemoglobin.
    E) histamine.
    Answer: A

41. The most abundant plasma protein is
    A) plasmin.
    B) alpha globulin.
    C) fibrinogen.
    D) albumin.
    E) beta globulin.
    Answer: D

42. The ion, which is essential in the blood-clotting mechanism, is
    A) magnesium.
    B) potassium.
    C) calcium.
    D) sodium.
    E) chlorine.
    Answer: C

43. Which of the following conditions is caused by the Epstein-Barr virus?
   A) polycythemia
   B) sickle-cell anemia
   C) iron-deficiency anemia
   D) septicemia
   E) infectious mononucleosis
   Answer: E

44. The percentage of red blood cells in whole blood is called the
   A) hemoglobin content.
   B) platelet count.
   C) buffy coat.
   D) hematocrit.
   E) serum content.
   Answer: D

45. All of the following are involved in blood clotting EXCEPT
   A) intrinsic factor.
   B) globulins.
   C) fibrinogen.
   D) prothrombin.
   E) calcium.
   Answer: B

## True-False

*Write T if the statement is true and F if the statement is false*

1. All blood cells originate from hemopoietic stem cells.
   Answer: True

2. The kidney excretes bilirubin.
   Answer: False

3. Neonatal jaundice disappears as the liver matures.
   Answer: True

4. Leukocytes can be divided into three major groups.
   Answer: False

5. Lymphocytes are the largest leukocytes.
   Answer: False

6. Monocytes that migrate to infected tissues are called wandering macrophages.
   Answer: True

7. Inflammation causes an increase in the hematocrit.
   Answer: False

8. Neutrophils are granulocytes.
   Answer: True

9. Plasma proteins are confined to blood.
   Answer: True

10. The extrinsic pathway of blood clotting is more complex than the intrinsic pathway.
    Answer: False

11. The breaking up of a blood clot is called clot retraction.
    Answer: False

12. Blood clotting always results in the formation of embolus.
    Answer: False

13. Heparin is produced by mast cells.
    Answer: True

14. A blood clot that is transported by the blood stream is called a thrombus.
    Answer: False

15. Agglutinins are also called isoantibodies.
    Answer: True

## Short Answer

*Write the word or phrase that best completes each statement or answers the question.*

1. Collective erythrocytes, leukocytes, and platelets are called the _____ elements of blood.
   Answer: formed

2. The branch of science concerned with the study of blood is called _____.
   Answer: hematology

3. The three functions of blood are transportation, regulation, and _____.
   Answer: protection

4. Blood belongs to _____ tissue.
   Answer: connective

5. Another name for immunoglobulins is _____.
   Answer: antibodies

6. The process of blood cell formation is called _____.
   Answer: hemopoiesis

7. The pigment that carries oxygen in red blood cells is _____.
   Answer: hemoglobin

8. The protein portion of hemoglobin is _____.
   Answer: globin

9. The urobilinogen reabsorbed into the bloodstream is converted into _____ that is excreted in urine.
   Answer: urobilin

10. Erythrocyte formation is called ____.
    Answer: erythropoisis

11. Leukocytes containing red or orange staining granules are called ____.
    Answer: eosinophils

12. Surface proteins on nucleated cells that can be used to identify tissues for transplantation are _____.
    Answer: major histocompatibility antigens

13. The function of platelets is ____.
    Answer: blood clotting

14. When platelets accumulate and attach to each other they form a mass called the ____.
    Answer: platelet plug

15. Breaking of blood clot is called ____.
    Answer: fibrinolysis

16. The surface of red blood cells contains antigens called ____.
    Answer: isoantigens or agglutinogens

17. Individuals with type A blood have ____ antibodies in the plasma.
    Answer: B

18. The Rh system of blood classification was first discovered in the blood of the ____.
    Answer: Rhesus monkey

19. The hemolysis caused by fetal-maternal incompatibility is called ____.
    Answer: erythroblastosis fetalis or hemolytic disease of the newborn

20. A condition in which the oxygen-carrying capacity to the blood is reduced is called ____.
    Answer: anemia

21. Bleeding, either internal or external is referred to as ____.
    Answer: hemorrhage

22. Septicemia is the medical term for ____.
    Answer: blood poisoning

23. The release of hemoglobin into the blood as a result of red blood cell rupture is called ____.
    Answer: hemolysis

24. During blood clotting prothrombin is converted into the enzyme ____.
    Answer: thrombin

25. Plasma minus the clotting proteins is called ____.
    Answer: serum

# Matching

*Choose the item from Column 2 that best matches each item in Column 1.*

1. Column 1: Contain hemoglobin.
   Column 2: erythrocytes

2. Column 1: Release lysozyme.
   Column 2: neutrophils

3. Column 1: Produce antibodies.
   Column 2: lymphocytes

4. Column 1: Become wandering macrophages.
   Column 2: monocytes

5. Column 1: Release histaminase
   Column 2: eosinophils

6. Column 1: Present in high numbers during parasitic infection.
   Column 1: eosinophils

7. Column 1: Known as mast cells.
   Column 2: basophils

8. Column 1: Caused by inadequate absorption or excessive loss of iron.
   Column 2: Iron-deficiency anemia

9. Column 1: Due to an insufficient production of erythrocytes.
   Column 2: Pernicious anemia

10. Column 1: An excessive loss of erythrocytes through bleeding.
    Column 2: Hemolytic anemia

11. Column 1: Characterized by distortion in the shape of erythrocytes.
    Column 2: Hemolytic anemia

12. Column 1: A result of the destruction or inhibition of red bond marrow.
    Column 2: Aplastic anemia

13. Column 1: Due to the production of abnormal hemoglobin.
    Column 2: Sickle-cell anemia

**Essay**

*Write your answer in the space provided or on a separate sheet of paper.*

1.  Describe the functions of blood.
    Answer:  Blood transports oxygen, carbon dioxide, nutrients, wastes, and hormones.  Blood helps to
    regulate pH, body temperature, and water content of the cells.  Furthermore, it prevents blood
    loss through clotting and also fights invading microbes.

2.  Name the different types of leukocytes.
    Answer:  1.  Granulocytes:  neutrophils, basophils, and eosinophils.
              2.  Agranulocytes:  lymphocytes and monocytes.

3.  Name the stages of blood clotting.
    Answer:  1.  Formation of prothrombinase (prothrombin activator).
              2.  Conversion of prothrombin into thrombin.
              3.  Conversion of soluble fibrinogen into insoluble fibrin.

## CHAPTER 15    The Cardiovascular System: Heart

## Multiple-Choice

*Choose the one alternative that best completes the statement or answers the question*

1. The outer portion of the pericardium is the
   A) epicardium.
   B) serous pericardium.
   C) fibrous pericardium.
   D) parietal pericardium.
   E) endocardium.
   Answer:  C

2. The layer of simple squamous epithelium that lines the inside of the myocardium is called
   A) epicardium.
   B) serous pericardium.
   C) fibrous pericardium.
   D) parietal pericardium.
   E) endocardium.
   Answer:  E

3. Cardiac muscle fibers are connected with each other by
   A) intercalated discs.
   B) tight junctions.
   C) desmosomes.
   D) neuromuscular junctions.
   E) synapses.
   Answer:  A

4. The portion of the heart that consists of cardiac muscle tissue is the
   A) epicardium.
   B) myocardium.
   C) pericardium.
   D) endocardium.
   E) All of the above.
   Answer:  B

5. Which portion of the heart has the thickest myocardium?
   A) left atrium
   B) right atrium
   C) left ventricle
   D) right ventricle
   E) they all have equal thickness
   Answer:  C

6. Blood transported by the pulmonary veins returns to the
   A) left atrium.
   B) left ventricle.
   C) right atrium.
   D) right ventricle.
   E) None of the above.
   Answer:  A

7. The muscular wall of the left ventricle is thicker than the right wall because it
   A) holds a greater volume of blood.
   B) pumps the blood to the lungs.
   C) pumps the blood to the entire body.
   D) pumps the blood through a smaller valve.
   E) pumps the blood through the heart.
   Answer: C

8. The groove separating the atria from the ventricles is known as
   A) interventricular septum.
   B) coronary sulcus.
   C) interventricular sulcus.
   D) interatrial septum.
   E) septum.
   Answer: B

9. The left ventricle pumps the blood into the
   A) pulmonary trunk.
   B) ascending aorta.
   C) descending aorta.
   D) pulmonary vein.
   E) vena cava.
   Answer: B

10. Which blood vessel delivers blood to the right atrium?
    A) vena cava.
    B) pulmonary artery.
    C) pulmonary trunk.
    D) pulmonary vein.
    E) aorta.
    Answer: A

11. The blood vessel that carries blood highest in oxygen is the
    A) ascending aorta.
    B) superior vena cava.
    C) pulmonary vein.
    D) coronary artery.
    E) pulmonary trunk.
    Answer: C

12. The valve between the left atrium and left ventricle is the
    A) bicuspid valve.
    B) tricuspid valve.
    C) aortic semilunar valve.
    D) pulmonary semilunar valve.
    E) sinoatrial valve.
    Answer: A

13. The valve between the left ventricle and the blood vessel leaving the left ventricle is the
    A) bicuspid valve.
    B) tricuspid valve.
    C) aortic semilunar valve.
    D) pulmonary semilunar valve.
    E) mitral valve.
    Answer: C

14. The blood vessel that collects deoxygenated blood of the coronary circulation and empties into the right atrium is the
    A) vena cava.
    B) sigmoid sinus.
    C) pulmonary sinus.
    D) coronary sinus.
    E) None of the above.
    Answer: D

15. The part of the circulation pumping blood to and from the lungs is known as
    A) systemic circulation.
    B) coronary circulation.
    C) pulmonary circulation.
    D) respiratory circulation.
    E) hepatic circulation.
    Answer: C

16. Death of an area of cardiac tissue due to an interrupted blood supply is called
    A) ischemia.
    B) myocardial infarction.
    C) angina pectoris.
    D) ventricular fibrillation.
    E) cardiac ischemia.
    Answer: B

17. The blood supply to the wall of the heart is supplied by the
    A) arch of the aorta.
    B) descending aorta.
    C) pulmonary artery.
    D) coronary sinus.
    E) coronary arteries.
    Answer: E

18. The normal pacemaker of the heart is/are the
    A) Purkinje fibers.
    B) bundle of His.
    C) atrioventricular node.
    D) sinoatrial node.
    E) atrioventricular bundle.
    Answer: D

19. Which of the following represents the correct sequence of structures in the cardiac conduction system?
   A) Purkinje fibers, AV nodes, SA node, bundle of His
   B) AV node, SA node, bundle of His, Purkinje fibers
   C) SA node, bundle of His, AV node, Purkinje fibers
   D) SA node, AV node, Purkinje fibers, bundle of His
   E) SA node, AV node, bundle of His, Purkinje fibers
   Answer: E

20. The P wave of an ECG indicates
   A) atrial depolarization.
   B) atrial repolarization.
   C) ventricular depolarization.
   D) ventricular repolarization.
   E) ventricular contraction.
   Answer: A

21. Atrial repolarization is indicated by the
   A) P wave.
   B) QRS wave.
   C) T wave.
   D) R portion of the QRS.
   E) None of the above.
   Answer: E

22. Which of the following represents the correct order of ECG waves?
   A) QRS, T, P
   B) QRS, P, T
   C) P, QRS, T
   D) T, QRS, P
   E) P, T, QRS
   Answer: C

23. At rest each cardiac cycle lasts approximately _____ seconds.
   A) 0.3
   B) 0.5
   C) 0.8
   D) 0.10
   E) 0.20
   Answer: C

24. The remaining 25% of the blood that fills the ventricles occurs during
   A) atrial diastole.
   B) atrial systole.
   C) ventricular diastole.
   D) ventricular systole.
   E) relaxation period.
   Answer: B

25. During ventricular filling
    A) the ventricles completely fill with blood before atrial contraction.
    B) ventricular contraction begins, and blood is pushed against the AV valves.
    C) the ventricles are about 75% filled with blood before the atria contract.
    D) the ventricles are starting to get filled with blood after atrial systole.
    E) None of the above.
    Answer: C

26. Closing of the AV valves produces the
    A) P wave.
    B) T wave.
    C) QRS wave.
    D) the first heart sound (lubb).
    E) the second heart sound (dubb).
    Answer: D

27. Heart murmurs are usually the result of defective
    A) nodes.
    B) valves.
    C) arteries.
    D) myocardial cells.
    E) Purkinje fibers.
    Answer: B

28. The cardiac output is
    A) the total volume of blood within the body.
    B) the milliliters of blood pumped per beat by each ventricle.
    C) the milliliters of blood pumped per minute by each ventricle.
    D) the number of beats per minute.
    E) the total amount of blood pumped by the heart per day.
    Answer: C

29. The cardiovascular center is located in the
    A) hypothalamus.
    B) thalamus.
    C) pons.
    D) midbrain.
    E) medulla oblongata.
    Answer: E

30. Which of the following causes an increase in the rate of the heartbeat?
    A) norepinephrine
    B) acetylcholine
    C) parasympathetic fibers
    D) Both A and C
    E) Both B and C
    Answer: A

31. Changes in the blood pressure are detected by
    A) osmoreceptors.
    B) baroreceptors.
    C) stretch receptors.
    D) chemoreceptors.
    E) proprioceptors.
    Answer: B

32. Impulses carried by means of the vagus nerve are
    A) parasympathetic impulses that increase the heart rate.
    B) parasympathetic impulses that decrease the heart rate.
    C) sympathetic impulses that increase the heart rate.
    D) sympathetic impulses that decrease the heart rate.
    E) autonomic impulses that both increase and decrease the heart rate.
    Answer: B

33. The average heartbeat of a healthy human is
    A) 60 beats per minute.
    B) 65 beats per minute.
    C) 75 beats per minute.
    D) 80 beats per minute.
    E) 85 beats per minute.
    Answer: C

34. The study of the heart and diseases associated with it is known as
    A) pathology.
    B) physiology.
    C) hematology.
    D) oncology.
    E) cardiology.
    Answer: E

35. The bulk of the heart consists of
    A) epicardium.
    B) endocardium.
    C) myocardium.
    D) serous pericardium.
    E) fibrous pericardium.
    Answer: C

36. The two upper chambers of the heart are separated by the
    A) interventricular septum.
    B) interatrial septum.
    C) parietal pericardium.
    D) visceral pericardium.
    E) coronary sulcus.
    Answer: B

37. Which of the following is/are risk factor(s) for the development of heart disease
    A) high blood pressure.
    B) diabetes mellitus.
    C) genetic predisposition.
    D) All of the above.
    E) None of the above.
    Answer: D

38. Excessive bleeding would
    A) decrease contraction strength.
    B) increase contraction strength.
    C) decrease the cardiac rate.
    D) have no affect of the cardiac rate.
    E) have no affect on contraction strength.
    Answer: B

39. The valve located between the right atrium and the right ventricle is the
    A) tricuspid valve.
    B) bicuspid valve.
    C) mitral valve.
    D) aortic valve.
    E) semilunar valve.
    Answer: A

40. When blood moves from an atrium to a ventricle
    A) the valve is pushed open.
    B) the papillary muscles relax.
    C) the corordae tendinae slacken.
    D) A and B are correct.
    E) All of the above.
    Answer: E

41. The blood flow through the numerous vessels in the myocardium is called
    A) portal circulation.
    B) pulmonary circulation.
    C) systemic circulation.
    D) hepatic circulation.
    E) coronary circulation.
    Answer: E

42. The medical term for heart attack is
    A) ischemia.
    B) angina pectoris.
    C) myocardial infarction.
    D) ventricular fibrillation.
    E) mitral valve prolapse.
    Answer: C

43. Cardiac output is measured in
    A) beats per minute.
    B) ml per minute.
    C) ml per beat.
    D) liters per beat.
    E) None of the above.
    Answer: B

44. Parasympathetic neurons reach the heart via the _____ cranial nerve.
    A) V
    B) VI
    C) VIII
    D) X
    E) XII
    Answer: D

45. A fluttering of the heart is called
    A) cardiac arrest.
    B) palpitation.
    C) paroxysmal tachycardia.
    D) ventricular fibrilation.
    E) cardiomegaly.
    Answer: B


## True-False

*Write T if the statement is true and F if the statement is false.*

1. The heart is located in the mediastenum.
   Answer: True

2. All arteries contain oxygenated blood, and all veins contain non-oxygenated blood.
   Answer: False

3. Atrioventricular valves are held closed by chordae tendinae and papillary muscles.
   Answer: True

4. The heart valves prevent the backflow of blood during the cardiac cycle.
   Answer: True

5. The autonomic nervous system initiates the contraction of the heart.
   Answer: False

6. Malfunctioning heart valves causes myocardial infarction.
   Answer: False

7. Blood flow through the heart is caused by changes in the size of the chambers.
   Answer: True

8. The atria receive all blood returning to the heart.
   Answer: True

9. Acetylcholine released by the parasympathetic fibers increases the rate of the heartbeat.
   Answer: False

10. Elevated levels of potassium decrease the heart rate and strength of the contraction.
    Answer: True

11. Mental states such as depression tend to increase the heart rate.
    Answer: False

12. Alcoholism may contribute to the development of heart disease.
    Answer: True

13. The rate of the heartbeat stays constant throughout life.
    Answer: False

14. Exercise increases cardiac efficiency and output.
    Answer: True

15. Overweight people develop extra capillaries to supply fat tissue.
    Answer: True

## Short Answer

*Write the word or phrase that best completes each statement or answers the question.*

1. The inner visceral layer of the pericardium is also called the ____.
   Answer: epicardium

2. The scientific study of the normal heart and the diseases associated with it is _____.
   Answer: cardiology

3. The space between the parietal and visceral pericardial membranes is the ____.
   Answer: pericardial cavity

4. Inflammation of the pericardium is called _____.
   Answer: pericarditis

5. The thickenings of the sarcolemma which connect the cardiac muscle fibers are ____.
   Answer: intercalated discs

6. The backflow of blood through an incompletely closed valve is called ____.
   Answer: regurgitation

7. The pouchlike structure on the anterior surface of each atrium is called _____.
   Answer: auricle

8. The right ventricle pumps blood into the ____.
   Answer: pulmonary trunk

9. The coronary arteries originate as branches of the ____.
   Answer: ascending aorta

10. The valve between the right atrium and the right ventricle is the _____ valve.
    Answer: tricuspid

11. The large vein in the back of the heart collecting deoxygenated blood is the ____.
    Answer: coronary sinus

12. The structure which picks up an action potential from the SA node is/are the ____.
    Answer: AV node

13. The medical term for heart attack is _____ _____.
    Answer: myocardial infarction

14. A recording of the electrical changes that accompany the heartbeat is called a(n) _____.
    Answer: electrocardiogram

15. The spread of the action potential through the ventricles is recorded as the ____ wave of an ECG.
    Answer: QRS

16. The specialized tissue capable of conducting action potentials and allowing the heart to beat without direct stimulus from the nervous system is the ____.
    Answer: conduction system

17. Contraction of the heart muscle is referred to as ____.
    Answer: systole

18. Heart sounds are produced by the ____ of the AV valves and the semilunar valves.
    Answer: closing

19. The amount of blood ejected by a ventricle during each contraction is called the ____.
    Answer: stroke volume

20. The ____ law of the heart explains the relationship between the stretching of the ventricular wall and the contraction strength.
    Answer: Starling's

21. A general term referring to an irregularity in the rhythm of the heart is ____.
    Answer: arrhythmia

22. A procedure that is used to visualize the coronary arteries, chambers, valves, and great vessels is ____.
    Answer: cardiac catheterization

23. The neurotransmitter released by the sympathetic fibers that increase the rate of the heartbeat is ____.
    Answer: norepinephrine

24. An incomplete closure of the interventricular septum causes ____.
    Answer: interventricular septal defect

25. A defect that exists at birth, and usually before, is called a(n) ____.
    Answer: congenital defect

## Matching

*Choose the item from Column 2 that best matches each item in Column 1.*

1. Column 1: The pacemaker of the heart.
   Column 2: SA node

2. Column 1: Located between the right atrium and the right ventricle.
   Column 2: tricuspid valve

3. Column 1: Located between the left atrium and the left ventricle.
   Column 2: bicuspid valve

4. Column 1: Emerge from the bundle branches.
   Column 2: Purkinje fibers

5. Column 1: Consists of cardiac muscle tissue.
   Column 2: myocardium

6. Column 1: The external layer of the heart wall.
   Column 2: epicardium

7. Column 1: Contraction of the heart chambers.
   Column 2: systole

8. Column 1: Relaxation of the heart chambers.
   Column 2: diastole

9. Column 1: Blood vessel associated with the left ventricle.
   Column 2: aorta

10. Column 1: Blood vessel associated with the right ventricle.
    Column 2: pulmonary trunk

## Essay

*Write your answer in the space provided or on a separate sheet of paper.*

1. Briefly describe the diagnostic value of an electrocardiogram.
Answer:  The ECG is used to diagnose abnormal cardiac rhythms and conduction patterns, detect the presence of life, and follow the course of recovery from a heart attack.

2. Name in correct sequence the structures involved as the cardiac impulse passes through the heart.
   Answer:  SA node, AV node, bundle of His, Purkinje fibers

3. Name the phases of the cardiac cycle.
   Answer:  The relaxation period, ventricular filling, and ventricular systole.

4. Define cardiac output.
   Answer:  Cardiac output (CO) is the amount of blood ejected by the left ventricle into the aorta per minute.  It is calculated as follow:  CO = stroke volume x beats per minute.

# CHAPTER 16    The Cardiovascular System: Blood Vessels and Circulation

## Multiple-Choice

*Choose the one alternative that best completes the statement or answers the question*

1.  Blood vessels that carry blood away from the heart are called
    A)  capillaries.
    B)  venules.
    C)  veins.
    D)  arteries.
    E)  All of the above.
    Answer:  D

2.  The smallest type of blood vessels are
    A)  arteries.
    B)  arterioles.
    C)  venules.
    D)  capillaries.
    E)  veins.
    Answer: D

3.  The endothelium consists of
    A)  simple cuboidal epithelium.
    B)  simple squamous epithelium.
    C)  epithelium and smooth muscle.
    D)  smooth muscle only.
    E)  epithelium, smooth muscle, and elastic fibers.
    Answer:  B

4.  The two main methods of capillary exchange are
    A)  diffusion and exocytosis.
    B)  endocytosis and exocytosis.
    C)  diffusion and bulk flow.
    D)  diffusion and filtration.
    E)  endocytosis and pinocytosis.
    Answer: C

5.  Exchange of nutrients and gases between the blood and tissue is the function of the
    A)  arteries.
    B)  arterioles.
    C)  veins.
    D)  venules.
    E)  capillaries.
    Answer:  E

6.  Which of the following statement(s) is/are true for bulk flow?
    A)  It is a passive transport mechanism.
    B)  It is an active transport mechanism.
    C)  It moves large numbers of molecules.
    D)  A and C
    E)  All of the above.
    Answer: D

7. Elasticity and contractility are properties of
   A) arteries.
   B) veins.
   C) capillaries.
   D) A and B.
   E) All of the above.
   Answer: A

8. Venous return is due to
   A) contraction of the left ventricle.
   B) skeletal muscle pump.
   C) respiratory pump.
   D) A only.
   E) A, B, and C.
   Answer: E

9. Blood vessels that are composed of a single layer of endothelial cells and a basement membrane are
   A) venules.
   B) arterioles.
   C) capillaries.
   D) Both A and B.
   E) None of the above.
   Answer: C

10. Venules
    A) are small veins.
    B) collect blood from capillaries.
    C) drain into veins.
    D) are similar in structure to arterioles.
    E) All of the above.
    Answer: E

11. Which of the following blood vessels are referred to as blood reservoirs
    A) arteries.
    B) veins.
    C) capillaries.
    D) All of the above.
    E) None of the above.
    Answer: B

12. Blood pressure is highest in the
    A) arteries.
    B) arterioles.
    C) veins.
    D) venules.
    E) capillaries.
    Answer: A

13. Resistance is related to
    A) blood viscosity.
    B) blood vessel length.
    C) blood vessel radius.
    D) All of the above.
    E) A and B only.
    Answer: D

14. Which of the following are involved in the regulation of blood pressure?
    A) baroreceptor reflexes
    B) chemoreceptor reflexes
    C) carotid bodies
    D) A and B
    E) A, B, and C
    Answer: E

15. Atrial natriuretic peptide
    A) is released by cells of the heart.
    B) lowers blood pressure.
    C) increases blood pressure.
    D) A and B.
    E) B and C.
    Answer: D

16. Which of the following is NOT a symptom of shock?
    A) decreased epinephrine levels
    B) increased levels of aldosterone
    C) rapid, resting heart rate
    D) cool, pale skin
    E) sweating
    Answer: A

17. Which of the following cells release chemicals that alter the diameter of blood vessels?
    A) macrophages
    B) white blood cells
    C) smooth muscle fibers
    D) endothelial cells
    E) all of the above
    Answer: E

18. Which of the following increase(s) blood pressure?
    A) increased cardiac rate
    B) increased peripheral resistance
    C) increased blood volume
    D) increased water retention
    E) All of the above
    Answer: E

19. Peripheral resistance
    A) increases as blood viscosity increases.
    B) decreases as blood viscosity increases.
    C) increases with increased blood vessel diameter.
    D) is higher in arteries than capillaries.
    E) decreases with increasing blood vessel length.
    Answer: A

20. The cardiovascular center is located in the
    A) hypothalamus.
    B) thalamus.
    C) medulla oblongata.
    D) midbrain.
    E) cerebral cortex.
    Answer: C

21. Sympathetic stimulation results in
    A) vasodilation.
    B) decrease in blood pressure.
    C) increase in blood pressure.
    D) A and B are correct.
    E) A and C are correct.
    Answer: C

22. Baroreceptors are located in the
    A) aorta.
    B) internal carotid arteries.
    C) jugular veins.
    D) in both A and B.
    E) in both B and C.
    Answer: D

23. Neurons that monitor carbon dioxide levels in the blood and are located in the carotid and aortic bodies are
    A) baroreceptors.
    B) chemoreceptors.
    C) mechanoreceptors.
    D) sympathetic neurons.
    E) parasympathetic neurons.
    Answer: B

24. Which of the following hormones influences blood pressure?
    A) ADH
    B) epinephrine
    C) norepinephrine
    D) renin
    E) all of these hormones influence blood pressure
    Answer: E

25. All of the following hormones cause an increase in blood pressure EXCEPT
    A) epinephrine.
    B) norepinephrine.
    C) renin.
    D) ANP.
    E) ADH.
    Answer: D

26. Which of the following substances can cause autoregulation of blood flow?
    A) prostaglandins
    B) nitric oxide
    C) carbon dioxide
    D) vasopressin
    E) ADH
    Answer: B

27. Which of the following blood vessels would have the greatest resistance to blood flow?
    A) a long, large diameter blood vessel
    B) a short, large diameter blood vessel
    C) a long, small diameter blood vessel
    D) a short, small diameter blood vessel
    E) a medium length, large diameter blood vessel
    Answer: C

28. The movement of blood from the abdominal veins into the thoracic veins is mostly due to
    A) pumping action of the heart.
    B) skeletal muscle contraction.
    C) pressure difference between the abdominal and thoracic cavity.
    D) the increasing blood pressure in the abdominal veins.
    E) All of the above.
    Answer: C

29. All of the following blood vessels are commonly used to feel the pulse EXCEPT
    A) radial artery.
    B) brachial artery.
    C) popliteal artery.
    D) femoral artery.
    E) common carotid artery.
    Answer: D

30. The pulse is a direct reflection of the
    A) cardiac output.
    B) blood pressure.
    C) venous return.
    D) cardiac rate.
    E) None of the above.
    Answer: D

31. The diastolic blood pressure
    A) provides information about the resistance of blood vessels.
    B) indicates the force of ventricular contraction.
    C) indicates the force of atrial contraction.
    D) is normally higher than the systolic blood pressure.
    E) indicates the cardiac output.
    Answer: A

32. Which of the following describes the pulmonary circulation?
    A) right atrium to left ventricle
    B) left ventricle to right atrium
    C) right ventricle to left atrium
    D) left atrium to right ventricle
    E) left ventricle to left atrium
    Answer: C

33. The blood vessel(s) containing blood with the highest oxygen contents is/are the
    A) pulmonary arteries.
    B) pulmonary veins.
    C) ascending aorta.
    D) descending aorta.
    E) thoracic aorta.
    Answer: B

34. All systemic blood vessels branch from the
    A) vena cava.
    B) pulmonary trunk.
    C) circle of Willis.
    D) aorta.
    E) None of the above.
    Answer: D

35. The circulatory route which bring blood to the tissues and back to the heart is the
    A) hepatic portal circulation.
    B) pulmonary circulation.
    C) aortic circulation.
    D) systemic circulation.
    E) cerebral circulation.
    Answer: D

36. The circle of Willis is part of the
    A) pulmonary circulation.
    B) cerebral circulation.
    C) hepatic portal circulation.
    D) systemic circulation.
    E) common circulation.
    Answer: B

37. The blood rich in substances absorbed from the gastrointestinal tract is carried by the
    A) fetal circulation.
    B) cerebral circulation.
    C) hepatic circulation.
    D) systemic circulation.
    E) pulmonary circulation.
    Answer: C

38. Which of the following are changes in the cardiovascular system related to the aging process?
    A) reduced cardiac output
    B) increased systolic blood pressure
    C) decreased maximum heart rate
    D) A and B
    E) A, B, and C
    Answer: E

39. The fetal circulation differs from the adult circulation because of two organ systems that are non-functional. These organ systems are the:
    A) cardiovascular and respiratory systems.
    B) cardiovascular and digestive systems.
    C) respiratory and digestive systems.
    D) nervous and respiratory systems.
    E) nervous and digestive systems.
    Answer: C

40. The coronary arteries branch from the
    A) arch of the aorta.
    B) thoracic artery.
    C) descending aorta.
    D) ascending aorta.
    E) None of the above.
    Answer: D

41. Which of the following arteries does NOT directly branch from the abdominal aorta?
    A) internal iliac
    B) common iliac
    C) celiac trunk
    D) superior mesenteric
    E) renal
    Answer: A

42. The direct continuation of the brachial artery is the _____ artery.
    A) ulnar
    B) axillary
    C) brachiocephalic
    D) radial
    E) digital
    Answer: D

43. The main vein of the heart is the
    A) superior vena cava.
    B) brachiocephalic vein.
    C) cardiac vein.
    D) coronary sulcus.
    E) coronary sinus.
    Answer: E

44. The external jugular veins empty into the
    A) dural venous sinuses.
    B) internal jugular veins.
    C) subclavian veins.
    D) brachiocephalic veins.
    E) superior vena cava.
    Answer: C

45. The formation of new blood vessels is referred to as:
    A) syncope.
    B) angiogenesis.
    C) occlusion.
    D) aortography.
    E) osteogenesis.
    Answer: B


## True-False

*Write T if the statement is true and F if the statement is false.*

1. Sympathetic stimulation causes vasoconstriction.
   Answer: True

2. All veins carry blood low in oxygen.
   Answer: False

3. Leaky valves cause varicose veins.
   Answer: True

4. Blood flow refers to the amount of blood that passes through a blood vessel in a given period of time.
   Answer: True

5. Blood always flows from regions of lower blood concentration to regions of higher blood concentration.
   Answer: False

6. Epinephrine has no effect on blood pressure.
   Answer: False

7. Peripheral resistance refers to the resistance to blood flow in the peripheral circulation.
   Answer: True

8. Blood volume only occasionally changes blood pressure.
   Answer: False

9. The vasomotor center is located in the hypothalamus.
   Answer: False

10. Parasympathetic stimulation via the vagus nerve causes a decrease in heart rate.
    Answer: True

11. Depression can cause a decrease in blood pressure.
    Answer: True

12. Muscle tissue is not dependent on autoregulation of blood flow.
    Answer: False

13. Velocity of blood flow depends on the cross-sectional area of the blood vessel.
    Answer: True

14. Skeletal muscle contractions help to return venous blood.
    Answer: True

15. Bradycardia indicates a rapid heart rate.
    Answer: False

## Short Answer

*Write the word or phrase that best completes each statement or answers the question.*

1. Arterioles within a tissue or organ branch into countless microscopic vessels called ____.
   Answer: capillaries

2. The movement of water and solutes out of capillaries into the interstitial fluid is called _____.
   Answer: filtration

3. The movement of water and solutes from the interstitial fluid into the capillaries is called _____.
   Answer: reabsorption

4. An increase in the size of a blood vessel lumen is referred to as ____.
   Answer: vasodilatation

5. The blood flow in capillaries is regulated by smooth muscle fibers called the ____.
   Answer: pericapillary sphincter

6. Chemoreceptors that are sensitive to carbon dioxide levels are found in the carotid and ____ bodies.
   Answer: aortic

7. Veins that lose their elasticity and become stretched and flabby are called _____ veins.
   Answer: varicose

8. Chemicals that alter the diameter of blood vessels are referred to as ____ factors.
   Answer: vasoactive

9. When hydrostatic pressure in capillaries is higher than osmotic pressure, ____ occurs.
   Answer: filtration

10. Fluid movement from interstitial fluid into capillaries is called ____.
    Answer: reabsorption

11. As blood flows from capillaries to venules to veins to the heart, its velocity ____.
    Answer: increases

12. The alternate expansion and recoiling of an artery with each contraction of the left ventricle is called ____.
    Answer: pulse

13. The instrument used to measure blood pressure is called ____.
    Answer: sphygmomanometer

14. Sweating during shock is due to ____ stimulation.
    Answer: sympathetic

15. The section between the diaphragm and the common iliac arteries is referred to as the ____.
    Answer: abdominal aorta

16. The blood vessel emerging from the right ventricle is the ____.
    Answer: pulmonary trunk

17. A dislodged thrombus which found its way into the pulmonary arterial blood flow causes ____.
    Answer: pulmonary embolism

17. A small artery is called a(n) ____.
    Answer: arteriole

18. Cardiac output is calculated by multiplying stroke volume by ____.
    Answer: heart rate

19. Blood flow from the left ventricle to the right atrium is the ____ circulation.
    Answer: systemic

20. The abdominal aorta branches into the ____.
    Answer: common iliac arteries

21. The cerebral circulation is a subdivision of the ____ circulation.
    Answer: systemic

22. All veins of the systemic circulation flow into either the superior or inferior vena cava or the ____.
    Answer: coronary sinus

23. The two pumps which help to return the blood to the heart are the skeletal muscle pump and the ____ pump.
    Answer: respiratory

24. Sympathetic impulses reach the heart via ____ ____ nerves.
    Answer: cardiac accelerator

25. The hormone released by cells of the kidneys in response to decreased blood volume or flow is ____.
    Answer: renin

## Matching

*Choose the item from Column 2 that best matches each item in Column 1.*

1. Column 1: Refers to rapid heart rate.
   Column 2: tachycardia

2. Column 1: Indicates a slow heart rate.
   Column 2: bradycardia

3. Column 1: Low blood pressure.
   Column 2: hypotension

4. Column 1: High blood pressure.
   Column 2: hypertension

5. Column 1: Inflammation of an artery.
   Column 2: arteritis

6. Column 1: Inflammation of a vein.
   Column 2: phlebitis

7. Column 1: Inflammation of a vein with clot formation.
   Column 2: thrombophlebitis

8. Column 1: The obstruction of a blood vessel lumen.
   Column 2: occlusion

9. Column 1: A temporary cessation of consciousness.
   Column 2: syncope

## Essay

*Write your answer in the space provided or on a separate sheet of paper.*

1. Describe the factors that affect blood pressure.
   Answer:  1. Any factor that increases cardiac output increases blood pressure.
   2. An increase in blood volume also increases blood pressure.
   3. Peripheral resistance to blood flow influences blood pressure.  Arterioles control peripheral resistance – and therefore blood pressure – by changing their diameters.

2. Name the factors that help venous return to the heart.
   Answer:  1. Pumping action of the heart.
   2. Velocity of blood flow.
   3. Skeletal muscle contractions.
   4. Valves in the veins.
   5. Breathing.

3. Briefly describe the systemic circulation.
   Answer:  Left ventricle – ascending aorta – arch of the aorta – descending aorta – thoracic aorta – abdominal aorta – organs and tissues – inferior and superior vena cava – right atrium.

## Multiple-Choice

*Choose the one alternative that best completes the statement or answers the question*

1.  Lymphatic tissue is a specialized form of
    A) loose connective tissue.
    B) reticular connective tissue.
    C) elastic connective tissue.
    D) epithelial tissue.
    E) glandular tissue.
    Answer: B

2.  All of the following belong to the lymphatic system EXCEPT
    A) lymph.
    B) lymphatic vessels.
    C) red bone marrow.
    D) yellow bone marrow.
    E) All of the above.
    Answer:  D

3.  Which of the following vitamins are transported by the lymphatic system?
    A) Vitamin A
    B) Vitamin E
    C) Vitamin C
    D) Both A and E
    E) Both B and C
    Answer: D

4.  All of the following organs contain lymphatic capillaries EXCEPT
    A) liver.
    B) central nervous system.
    C) peripheral nervous system.
    D) intestine.
    E) skin.
    Answer: B

5.  The thoracic duct
    A) receives lymph from the left side of the head.
    B) receives lymph from the upper right side of the body.
    C) empties the lymph into the right subclavian vein.
    D) both B and C are correct.
    E) both A and C are correct.
    Answer: A

6.  Which of the following cells produce antibodies
    A) T-lymphocytes.
    B) B-lymphocytes.
    C) monocytes.
    D) phagocytes.
    E) erythrocytes.
    Answer:  B

7. Which of the following is NOT surrounded by a connective tissue capsule?
    A) spleen
    B) thymus gland
    C) lymph nodes
    D) lymphatic nodules
    E) all of the above are surround by a capsule
    Answer: D

8. The lymphatic system
    A) transports lipids from the gastrointestinal tract to the blood.
    B) returns filtered proteins to the cardiovascular system.
    C) returns excess interstitial fluid to the cardiovascular system.
    D) functions in surveillance and defense.
    E) All of the above are correct.
    Answer: E

9. Lymphatic vessels
    A) resemble arteries.
    B) resemble arterioles.
    C) resemble veins.
    D) have less valves than veins.
    E) return lymph directly to the heart.
    Answer: C

10. Which of the following is that largest single mass or lymphatic tissue?
    A) tonsils
    B) spleen
    C) lymph nodes
    D) bone marrow
    E) lymphatic nodules
    Answer: B

11. Edema can be a result of
    A) an infected lymph node.
    B) blockage of a lymphatic vessel.
    C) increased blood pressure.
    D) A and B are correct.
    E) A, B, and C are correct.
    Answer: E

12. Lymph nodes
    A) are bean-shaped organs.
    B) are located along lymphatic vessels.
    C) are divided into follicles.
    D) are scattered throughout the body.
    E) All of the above.
    Answer: E

13. Afferent lymphatic vessels are characteristic for
    A) tonsils.
    B) the spleen.
    C) lymphatic nodules.
    D) lymph nodes.
    E) the thymus gland.
    Answer: D

14. Worn-out and damaged red blood cells are destroyed in the
    A) thymus gland.
    B) tonsils.
    C) spleen.
    D) lymph nodes.
    E) lymphatic vessels.
    Answer: C

15. Hemopoiesis during fetal life occurs in the
    A) liver.
    B) spleen.
    C) thymus gland.
    D) lymph nodes.
    E) None of the above are correct.
    Answer: B

16. T-cell maturation is a function of the
    A) liver.
    B) spleen.
    C) tonsils.
    D) thymus.
    E) All of the above.
    Answer: D

17. Which of the following is/are characteristic structure(s) of the spleen?
    A) efferent lymphatic vessels
    B) afferent lymphatic vessels
    C) T-lymphocytes
    D) red pulp
    E) Both A and B.
    Answer: D

18. The efferent vessels from the last nodes in a chain unite to form
    A) the thoracic duct.
    B) left subclavian vein.
    C) the right lymphatic duct.
    D) the main lymphatic vein.
    E) a lymph trunk.
    Answer: E

19. Which of the following carry out the function of the spleen after splenectomy?
    A) thymus
    B) lymph nodes
    C) liver
    D) red bone marrow
    E) both C and D
    Answer: E

20. The first line of defense against pathogens is
    A) phagocytosis.
    B) production of antibodies.
    C) inflammation.
    D) the intact skin.
    E) immunity.
    Answer: D

21. All of the following are characteristic for the second line of defense EXCEPT:
    A) phagocytes.
    B) sebum.
    C) antimicrobial proteins.
    D) inflammation.
    E) fever.
    Answer: B

22. Lysozyme is found in all of the following EXCEPT
    A) tears.
    B) saliva.
    C) vaginal secretions.
    D) nasal secretions.
    E) tissue fluids.
    Answer: C

23. Which of the following is considered a chemical factor in the first line of defense
    A) flow of urine.
    B) mucous membrane.
    C) saliva.
    D) vomiting.
    E) gastric juice.
    Answer: E

24. Complement proteins
    A) are found in blood plasma.
    B) are present in infected cells.
    C) are produced by T-cells.
    D) are interferons.
    E) are produced by B-lymphocytes.
    Answer: A

25. Macrophages develop from
    A) lymphocytes.
    B) monocytes.
    C) neutrophils.
    D) basophils.
    E) eosinophils.
    Answer: B

26. Which of the following represents the correct order in phagocytosis?
    A) chemotaxis, ingestion, adherence
    B) adherence, chemotaxis, ingestion
    C) ingestion, chemotaxis, adherence
    D) chemotaxis, adherence, ingestion
    E) none of the above
    Answer: D

27. Pseudopods are formed during
    A) ingestion.
    B) adherence.
    C) chemotaxis.
    D) All of the above.
    E) None of the above.
    Answer: A

28. Which of the following organs does NOT contain fixed macrophages?
    A) lungs
    B) brain
    C) liver
    D) spleen
    E) all of the above contain fixed macrophages
    Answer: E

29. All of the following are symptoms of inflammation EXCEPT
    A) pain.
    B) redness.
    C) fever.
    D) swelling.
    E) heat.
    Answer: C

30. The first stage in the inflammatory response is
    A) redness.
    B) fever.
    C) vasoconstriction.
    D) vasodilation.
    E) phagocyte migration.
    Answer: D

31. Which of the following can act as an antigen?
    A) bacteria
    B) viruses
    C) pollen
    D) food
    E) All of the above.
    Answer: E

32. Vaccination is an example of
    A) naturally acquired active immunity.
    B) naturally acquired passive immunity.
    C) artificially acquired active immunity.
    D) artificially acquired passive immunity.
    E) None of the above.
    Answer: C

33. Mother's milk will provide the infant with
    A) naturally acquired active immunity.
    B) naturally acquired passive immunity.
    C) artificially acquired active immunity.
    D) artificially acquired passive immunity.
    E) lifelong immunity.
    Answer: B

34. Cell mediated immunity is provided by
    A) phagocytes.
    B) macrophages.
    C) basophils.
    D) T-cells.
    E) B-cells.
    Answer: D

35. Which of the following stimulate the development of B-cells into antibody-producing plasma cells?
    A) suppressor T-cells
    B) cytotoxic T-cells
    C) helper T-cells
    D) memory T-cells
    E) delayed hypersensitivity cells
    Answer: C

36. The cells that are important in preventing autoimmune disease are
    A) memory T-cells.
    B) helper T-cells.
    C) plasma cells.
    D) suppressor T-cells.
    E) memory B-cells.
    Answer: D

37. The first antibody present at an initial contact with an antigen is
    A) IgA.
    B) IgG.
    C) IgE.
    D) IgD.
    E) IgM.
    Answer: E

38. The most abundant antibody in the circulation is
    A) IgA.
    B) IgG.
    C) IgE.
    D) IgD.
    E) IgM.
    Answer: B

39. Which of the following antibodies is associated with hypersensitivity reactions?
    A) IgA
    B) IgG
    C) IgE
    D) IgD
    E) IgM
    Answer: C

40. A cell infected by a virus will produce
    A) antibodies.
    B) immunoglobulins.
    C) prostaglandins.
    D) interferons.
    E) interleukin.
    Answer: D

41. Immunologic surveillance is carried out by
    A) B-cells.
    B) cytotoxic T-cells.
    C) memory T-cells.
    D) helper T-cells.
    E) suppressory T-cells.
    Answer: B

42. The initial response to HIV invasion is a modest decline in the number of circulating
    A) monocytes.
    B) B-lymphocytes.
    C) cytotoxic T-cells.
    D) memory T-cells.
    E) helper T-cells.
    Answer: E

43. Which of the following is NOT an autoimmune disease?
    A) rheumatoid arthritis
    B) Hodgkin's disease
    C) Addison's disease
    D) rheumatic fever
    E) multiple sclerosis
    Answer: B

44. Before detection the AIDS virus can be present and proliferating in
    A) macrophages.
    B) helper T-cells.
    C) the blood.
    D) skin cells.
    E) B-cells.
    Answer: A

45. The Epstein-Barr virus causes
    A) Addison's disease.
    B) Hodgkin's disease.
    C) infectious mononucleosis.
    D) AIDS.
    E) multiple sclerosis.
    Answer: C


## True-False

*Write T if the statement is true and F if the statement is false.*

1. When fluid bathes the cells, it is called interstitial fluid, when it flows through lymphatic vessels it is called lymph.
   Answer: True

2. The thymus gland is located in the thoracic cavity.
   Answer: True

3. Edema is an accumulation of blood in tissue spaces.
   Answer: False

4. Immunity is the production of specific antibodies against a specific pathogen or toxin.
   Answer: True

5. Body cells infected with bacteria produce proteins called interferons.
   Answer: False

6. Phagocytosis is the ingestion of microbes or any foreign particles or cell debris by phagocytes.
   Answer: True

7. Macrophages that remain in certain tissues and organs of the body are called fixed macrophages.
   Answer: True

8. Only severe inflammations produce pus.
   Answer: False

9. Only microbes are considered antigens in the human body.
   Answer: False

10. Before T cells can attack antigens, the T cells must become sensitized.
    Answer: True

11. Cytotoxic T cells destroy target cells on contact.
    Answer: True

12. Delayed hypersensitivity cells are activated in response to allergic reactions.
    Answer: True

13. Memory B cells produce antibodies.
    Answer: False

14. When a normal cell transforms into a cancer cell, it may display cell surface components called tumor antigens.
    Answer: True

15. It has been shown that mosquitoes can transmit the HIV virus.
    Answer: False

## Short Answer

*Write the word or phrase that best completes each statement or answers the question.*

1. Lymphatic vessels lymph to the venous blood in the _____ veins.
   Answer: subclavian

2. Lymph leaves a lymph node through the _____ lymphatic vessel.
   Answer:  efferent

3. The main collecting duct of the lymphatic system that receives lymph from the left side of the body is the _____ _____.
   Answer:  thoracic duct

4. The "milking action" of skeletal muscle contractions is referred to as the _____ _____ _____.
   Answer: skeletal muscle pump

5. The spread of cancer cells from a primary tumor to other sites of the body with the help of the lymphatic system is referred to as _____.
   Answer: metastasis

6. The term MALT stands for _____-_____ _____ _____.
   Answer: mucosa-associated lymphatic tissue

7. Disease producing organisms are called _____.
   Answer:  pathogens

8. The ability to ward off disease is called _____.
   Answer:  resistance

9. Defecation and vomiting are examples of the _____ line of defense.
   Answer: first

10. The lack of resistance is called ____.
    Answer: susceptibility

11. Some complement proteins produce holes in the plasma membrane of microbes causing the contents of the microbes to leak out by a process called ____.
    Answer: cytolysis

12. When an infection occurs, neutrophils and ____ migrate to the infected area.
    Answer: monocytes

13. Materials indigestible by phagocytes remain in cellular structures called ____.
    Answer: residual bodies

14. The injection of antibodies from an outside source is an example of ____ immunity.
    Answer: artificially acquired passive

15. A protein produced by plasma cells in response to the presence of a foreign substance is called ____.
    Answer: antibody

16. Cells believed to depress parts of the immune response are called ____.
    Answer: suppressor T-cells

17. During a secondary immune response plasma cells produce mainly ____ antibodies.
    Answer: IgG

18. Despite immunologic surveillance, some cancer cells escape destruction, a phenomenon called ____.
    Answer: immunologic escape

19. Infections that take advantage of a person's compromised immune system are called ____ infections.
    Answer: opportunistic

20. A person who is overly reactive to an antigen is said to be ____.
    Answer: hypersensitive (allergic)

21. The antigens that induce an allergic reaction are called ____.
    Answer: allergens

22. The removal of a lymph node is called ____.
    Answer: lymphadectomy

23. A tumor composed of lymphatic tissue is called ____.
    Answer: lymphoma

24. The replacement of an injured or diseased tissue or organ is called ____.
    Answer: transplantation

25. Iron-binding proteins that inhibit the growth of certain bacteria are called _____.
    Answer: transferrins

## Matching

*Choose the item from Column 2 that best matches each item in Column 1.*

1. Column 1: A form of cancer usually arising in lymph nodes.
   Column 2: Hodgkin's disease

2. Column 1: An autoimmune, noncontagious, inflammatory disease of connective tissue.
   Column 2: Systemic lupus erythematosus

3. Column 1: A condition suffered by the "bubble boy".
   Column 2: Severe combined immunodeficiency

4. Column 1: Enlarged, tender, and inflamed lymph nodes.
   Column 2: Adenitis

5. Column 1: Inflammation of lymphatic vessels.
   Column 2: Lymphangitis

6. Column 1: A benign tumor of the lymphatic vessels.
   Column 2: lymphangioma

7. Column 1: Ingestion and destruction of microbes.
   Column 2: phagocytosis

8. Column 1: Engulfing a microbe into a phagocytotic vesicle.
   Column 2: ingestion

9. Column 1: Attachment of macrophage to a microbe.
   Column 2: adherence

10. Column 1: Promotion of phagocytosis by coating with a complement.
    Column 2: opsonization

## Essay

*Write your answer in the space provided or on a separate sheet of paper.*

1. Name the functions of the lymphatic system.
   Answer:  1. Draining intestinal fluid.
   2. Transporting lipids and lipid-soluble vitamins from the gastrointestinal tract to the blood.
   3. Protection of the body from foreign cells, microbes, and cancer cells.

2. List the mechanical factors in the first line of defense against disease causing microorganisms.
   Answer:  skin, mucous membranes, lacrimal apparatus, saliva, mucus, hairs, cilia, epiglottis, flow of urine, defecation, vomiting.

3. Describe the phases of phagocytosis.

    Answer:    1. Chemotaxis is the chemical attraction of phagocytes to microorganisms. These chemicals might come from the invading microbes, white blood cells. Damaged tissue cells, or activated complement.

    2. Adherence, the phagocyte attaches itself to the microorganism or other foreign material.

    3. Ingestion, the phagocyte engulfs the microorganism or foreign substance with the help of pseudopods.

## CHAPTER 18   The Respiratory System

## Multiple-Choice

*Choose the one alternative that best completes the statement or answers the question*

1.  Which of the following belongs to the respiratory portion of the lower respiratory system?
    A)  larynx
    B)  trachea
    C)  terminal bronchioles
    D)  alveolar ducts
    E)  both C and D
    Answer: D

2.  The exchange of gases between blood and cells is called
    A)  pulmonary ventilation.
    B)  internal respiration.
    C)  external respiration.
    D)  cellular respiration.
    E)  inspiration.
    Answer:  B

3.  Which of the following does NOT belong to the conducting portion of the respiratory system?
    A)  alveolar ducts
    B)  terminal bronchioles
    C)  bronchioles
    D)  nose
    E)  pharynx
    Answer:  A

4.  The internal nose is connected to the pharynx through the
    A)  nasal septum.
    B)  nasal conchae.
    C)  paranasal sinuses.
    D)  external nares.
    E)  internal nares.
    Answer:  E

5.  The nasal cavity is lined with
    A)  simple columnar epithelium.
    B)  transitional epithelium.
    C)  pseudostratified ciliated columnar epithelium.
    D)  stratified ciliated columnar epithelium.
    E)  simple ciliated cuboidal epithelium.
    Answer:  C

6.  The palatine tonsils are found in the
    A)  nasopharynx.
    B)  oropharynx.
    C)  laryngopharynx.
    D)  larynx.
    E)  None of the above.
    Answer: B

7. The structure with openings to the Eustachian tubes is the
   A) larynx.
   B) fauces.
   C) oropharynx.
   D) nasopharynx.
   E) laryngopharynx.
   Answer: D

8. The nasal cavity is divided into right and left sides by the
   A) external nares.
   B) internal nares.
   C) nasal conchae.
   D) nasal septum.
   E) paranasal sinuses.
   Answer: D

9. Which of the following is called the Adam's apple?
   A) thyroid cartilage
   B) epiglottis
   C) cricoid cartilage
   D) arytenoids cartilage
   E) glottis
   Answer: A

10. The structure which closes off the larynx is the
    A) glottis.
    B) Adam's apple.
    C) epiglottis.
    D) tongue.
    E) vocal cords.
    Answer: C

11. The greater the pressure of air against the vocal cords,
    A) the higher the pitch.
    B) the lower the sound.
    C) the louder the sound.
    D) the lower the pitch.
    E) the hoarser the sound.
    Answer: C

12. The false vocal cords
    A) aid in the production of sound.
    B) are located in the lower larynx.
    C) produce high pitch sounds.
    D) All of the above.
    E) None of the above.
    Answer: E

13. Which of the following describes a correct order of structures in the respiratory passageways?
    A) pharynx, trachea, larynx, bronchi, bronchioles
    B) larynx, pharynx, trachea, bronchioles, bronchi
    C) trachea, pharynx, larynx, bronchi, bronchioles
    D) pharynx, larynx, trachea, bronchi, bronchioles
    E) pharynx, larynx, trachea, bronchioles, bronchi
    Answer: D

14. Terminal bronchioles are lined with
    A) pseudostratified ciliated columnar epithelium.
    B) simple cuboidal epithelium.
    C) simple ciliated columnar epithelium.
    D) simple squamous epithelium.
    E) simple ciliated squamous epithelium.
    Answer: D

15. The C-shaped rings that provide support for the wall of the trachea are made out of
    A) hyaline cartilage.
    B) elastic cartilage.
    C) fibrocartilage.
    D) reticular connective tissue.
    E) bone.
    Answer: A

16. Histamine
    A) has no effect on bronchioles.
    B) causes bronchiole constriction.
    C) causes bronchiole dilation.
    D) causes muscle spasm.
    E) causes inflammation of the bronchioles.
    Answer: B

17. Epinephrine
    A) has no effect on bronchioles.
    B) causes bronchiole constriction.
    C) causes bronchiole dilation.
    D) causes muscle spasm.
    E) causes inflammation of the bronchioles.
    Answer: C

18. The pain in the early stages of pleurisy is caused by
    A) bronchiole constriction.
    B) bronchiole dilation.
    C) friction between the swollen membranes.
    D) histamine release.
    E) parasympathetic stimulation.
    Answer: C

19. Which of the following is NOT a structure associated with the lungs?
    A) visceral pleura
    B) parietal pleura
    C) cardiac notch
    D) endocardium
    E) hilus
    Answer: D

20. The exchange of gases occurs in the
    A) trachea.
    B) terminal bronchioles.
    C) alveoli.
    D) primary bronchus.
    E) All of the above.
    Answer: C

21. Surfactant is produced by
    A) terminal bronchioles.
    B) alveolar macrophages.
    C) squamous pulmonary epithelial cells.
    D) septal cells.
    E) capillary membranes.
    Answer: D

22. Which of the following divides into alveolar ducts?
    A) secondary bronchioles
    B) respiratory bronchioles
    C) tertiary bronchioles
    D) terminal bronchioles
    E) alveoli
    Answer: B

23. For air to enter the lungs during inspiration
    A) the pressure inside the lungs must become lower than the atmospheric pressure.
    B) the pressure inside the lungs must be higher than the atmospheric pressure.
    C) the pressure inside the lungs must be equal to the atmospheric pressure.
    D) the size of the lungs must be decreased.
    E) the diaphragm has to be relaxed.
    Answer: A

24. The lungs contain about _____ alveoli.
    A) 10,000
    B) 300,000
    C) one million
    D) 300 million
    E) 500 million
    Answer: D

25. The volume of air that can be exhaled after normal exhalation is the
   A) tidal volume.
   B) residual volume.
   C) inspiratory reserve volume.
   D) expiratory reserve volume.
   E) minute volume of respiration.
   Answer: D

26. The volume of air in a normal breath is called
   A) total lung capacity.
   B) vital capacity.
   C) tidal volume.
   D) functional residual capacity.
   E) residual volume.
   Answer: C

27. The total lung capacity equals
   A) TV + IRV + ERV + RV.
   B) IRV + ERV + RV.
   C) TV + IRV.
   D) FRC + ERV + RV.
   E) TV + ERV + IRV.
   Answer: A

28. Gas exchange in the lungs happens by the process of
   A) osmosis.
   B) diffusion.
   C) exocytosis.
   D) active transport.
   E) filtration.
   Answer: B

29. All of the following decrease the efficiency of external respiration EXCEPT
   A) morphine.
   B) increased alveolar $PO_2$.
   C) increased altitude.
   D) pneumonia.
   E) high alveolar $PCO_2$.
   Answer: B

30. Most oxygen in the blood is transported
   A) as deoxyhemoglobin.
   B) as gas dissolved in plasma.
   C) as oxyhemoglobin.
   D) as carboxyhemoglobin.
   E) as reduced hemoglobin.
   Answer: C

31. Nerve impulses travel from the active inspiratory area to the diaphragm via the
    A) vagus nerve.
    B) phrenic nerve.
    C) thoracic nerve.
    D) intercostals nerve.
    E) radial nerve.
    Answer: B

32. The pneumotaxic area is located in the
    A) medulla oblongata.
    B) midbrain.
    C) hypothalamus.
    D) pons.
    E) cerebellum.
    Answer: D

33. Approximately how much $CO_2$ in the blood is carried as bicarbonate?
    A) 40%
    B) 50%
    C) 60%
    D) 70%
    E) 80%
    Answer: D

34. When stretch receptors in the lungs are activated
    A) expiration will occur.
    B) the lungs will deflate.
    C) the apneustic area are inhibited.
    D) impulses are sent along the vagus nerve.
    E) All of the above.
    Answer: E

35. The primary chemical stimulus for breathing is the concentration of
    A) carbon monoxide in the blood.
    B) $H^+$ and carbon dioxide concentration of blood.
    C) the carbon dioxide concentration of the blood.
    D) the oxygen concentration of the blood.
    E) the carbonic acid concentration in the blood.
    Answer: B

36. Which of the following is NOT a factor in the control of respiration?
    A) a drop in blood pressure
    B) increased blood pressure
    C) pain
    D) None of the above influence respiration
    E) All of the above influence respiration
    Answer: E

37. The basic rhythm of respiration is controlled by the
    A) pons.
    B) medulla oblongata.
    C) hypothalamus.
    D) pneumotaxic area.
    E) apneustic area.
    Answer: B

38. Painful or labored breathing is referred to as
    A) rhinitis.
    B) diptheria.
    C) dyspnea.
    D) orthopena.
    E) epistaxis.
    Answer:  C

39. A condition in infants that is due to a lack of surfactant is
    A) pneumonia.
    B) sudden infant death syndrome.
    C) respiratory distress syndrome.
    D) pulmonary edema.
    E) asthma.
    Answer:  C

40. A disorder characterized by the destruction of the alveolar walls is
    A) chronic bronchitis.
    B) emphysema.
    C) chronic obstructive pulmonary disease.
    D) asthma.
    E) tuberculosis.
    Answer: B

## True-False

*Write T if the statement is true and F if the statement is false.*

1. The first step of respiration is external respiration.
   Answer: False

2. The nostrils are also called external nares.
   Answer: True

3. The middle portion of the pharynx is the nasopharynx.
   Answer: False

4. The mucous membrane of the larynx forms two pairs of folds.
   Answer: True

5. The trachea is located lateral to the esophagus.
   Answer: False

6. Tertiary bronchi divide into terminal bronchioles.
   Answer: False

7. The narrow top portion of the lung is called the apex.
   Answer: True

8. The right lung is divided into three lobes.
   Answer: True

9. Oxygenated blood for the lungs own tissues is delivered through the pulmonary arteries.
   Answer: False

10. In order for respiration to occur, the volume of the lung needs to be increased.
    Answer: True

11. The pressure inside the lungs is the alveolar pressure.
    Answer: True

12. The record of pulmonary volumes and capacities is called a spirogram.
    Answer: True

13. In clinical practice the word ventilation means inspiration.
    Answer: False

14. The total pressure of a gas mixture is calculated by multiplying the partial pressures.
    Answer: False

15. The transport of respiratory gases between the lungs and body tissues is a function of the blood.
    Answer: True

## Short Answer

*Write the word or phrase that best completes each statement or answers the question.*

1. The lowest portion of the pharynx is the _____.
   Answer: laryngopharynx

2. The branches of the trachea to the bronchi and bronchioles is referred to as the _____.
   Answer: bronchial tree

3. The membrane that encloses and protects the lungs is the _____.
   Answer: pleural membrane

4. Each lobe of the lungs is divided into regions called _____.
   Answer: bronchopulmonary segments

5. Terminal bronchioles subdivide into microscopic branches called _____.
   Answer: respiratory bronchioles

6. The maneuver used to expel an aspirated object is called the _____ maneuver.
   Answer: Heimlich

7. The respiratory gases move across the _____ membrane.
   Answer: alveolar-capillary (respiratory)

8. The visual examination of bronchi through a bronchoscope is called _____.
   Answer: bronchoscopy

9. When the diaphragm contracts it _____.
   Answer: flattens

10. The term applied to normal quiet breathing is _____.
    Answer: eupnea

11. The phospholipids produced by the septal cells are called _____.
    Answer: surfactant

12. The total volume of air taken in during one minute is called the _____.
    Answer: minute volume of respiration

13. The air that remains in the lungs after the expiratory reserve volume is expelled, is the _____.
    Answer: residual volume

14. The sum of residual volume plus expiratory reserve volume is the _____.
    Answer: functional residual capacity

15. The _____ area controls the basic rhythm of respiration.
Answer: medullary rhythmicity

16. The passive process by which air flows into and out of the lungs is called _____.
    Answer: ventilation (breathing)

17. The apneustic area is located in the _____.
    Answer: lower pons

18. The protective mechanism that prevents overinflation of the lungs is called _____.
    Answer: inflation reflex

19. A slow rate and depth of respiration is called _____.
    Answer: hypoventilation

20. The temporary cessation of breathing is known as _____.
    Answer: apnea

21. The structure that prevents food from entering the respiratory passages is the _____.
    Answer: epiglottis

22. The chemosensitive area is located in the _____.
    Answer: medulla oblongata

23. Carbon dioxide can be carried by hemoglobin as _____.
    Answer: carbaminohemoglobin

24. The immediate increase in ventilation at the onset of exercise is a result of the stimulation of _____.
    Answer: proprioceptors

25. A chronic or acute inflammation of the mucous membranes in the nose is called _____.
    Answer: rhinitis

## Matching

*Choose the item from Column 2 that matches each item in Column 1.*

1. Column 1:  The alveolar walls lose their elasticity and remain filled with air during expiration.
   Column 2:  emphysema

2. Column 1:  Also called hyaline membrane disease
   Column 2:  respiratory distress syndrome

3. Column 1:  Kills infants between the ages of 1 week and 12 months, without warning
   Column 2:  sudden infant death syndrome

4. Column 1:  The presence of a blood clot in a pulmonary arterial vessel
   Column 2:  pulmonary embolism

5. Column 1:  An abnormal accumulation of interstitial fluid in the interstitial spaces and alveoli of the lungs
   Column 2:  pulmonary edema

6. Column 1:  Another term for nosebleed
   Column 2:  epistaxis

7. Column 1:  Spitting of blood from the respiratory tract
   Column 2:  hemoptysis

8. Column 1:  Oxygen starvation
   Column 2:  asphyxia

9. Column 1:  Painful breathing
   Column 2:  dyspnea

10. Column 1:  Difficult breathing that occurs in the horizontal position
    Column 2:  orthopnea

## Essay

*Write your answer in the space provided or on a separate sheet of paper.*

1. Name, in correct order, the conducting portion of the respiratory system.
   Answer:  nose, pharynx, larynx, trachea, bronchi, bronchioles, terminal bronchioles

2. Name the function of the interior structures of the nose.
   Answer:    1. They warm, moisten, and filter incoming air.
   2.  They receive olfactory stimuli.
   3.  They provide a resonating chamber for speech sounds.

3. Name and briefly describe the three basic processes of respiration.

   Answer:   1. Pulmonary ventilation is the movement of air in and out of the lungs, inspiration occurs when the pressure inside the lungs is lower than the atmospheric pressure. This is achieved by increasing the volume of the lungs. Expiration occurs when the pressure in the lungs is greater than the pressure of the atmosphere. This occurs when the size of the thoracic cavity decreases.

   2. External respiration is the exchange of gases between the air in the alveoli and the blood.

   3. Internal respiration is the exchange of gases between the blood and tissues.

4. Define asthma.

   Answer:   Asthma is a reaction, usually allergic, characterized by attacks of wheezing and difficult breathing, spasms of the smooth muscles in the walls of smaller bronchi and bronchioles cause the passageways to close partially, resulting in an attack.

## CHAPTER 19    THE DIGESTIVE SYSTEM

## Multiple-Choice

*Choose the one alternative that best completes the statement or answers the question.*

1.  All of the following are part of the gastrointestinal tract except the
    A) stomach.
    B) gallbladder.
    C) esophagus.
    D) small intestine.
    E) pharynx.
    Answer: B

2.  Which of the following is NOT an accessory structure of the digestive system?
    A) liver
    B) gallbladder
    C) pancreas
    D) spleen
    E) teeth
    Answer: D

3.  The layer of the GI tract wall that contains blood and lymphatic vessels is the
    A) muscularis mucosa.
    B) mucosa.
    C) submucosa.
    D) serosa.
    E) muscularis.
    Answer: C

4.  The ability of the GI tract to mix and move material along its length is called
    A) digestion.
    B) peristalsis.
    C) absorption.
    D) motility.
    E) excitability.
    Answer: D

5.  The greater omentum is part of the
    A) mucosa.
    B) liver.
    C) mesentery.
    D) peritoneum.
    E) large intestine.
    Answer: D

6.  Enteric neurons are found in the
    A) mucosa.
    B) submucosa.
    C) muscularis.
    D) serosa.
    E) Both A and B.
    Answer: B

7. The projection hanging from the soft palate is/are the
   A) papillae.
   B) uvula.
   C) lingual frenulum.
   D) fauces.
   E) tonsils.
   Answer: B

8. Which of the following contains taste buds?
   A) papillae
   B) uvula
   C) tonsils
   D) fauces
   E) none of the above
   Answer: A

9. Mumps is an inflammation of the
   A) sublingual glands.
   B) submandibular glands.
   C) parotid glands.
   D) thyroid glands.
   E) parathyroid glands.
   Answer: C

10. The tongue is composed of
    A) smooth muscle.
    B) skeletal muscle.
    C) loose connective tissue.
    D) glandular tissue.
    E) None of the above.
    Answer: B

11. Salivary amylase secreted into the oral cavity starts the digestion of
    A) proteins.
    B) starch.
    C) lipids.
    D) amino acids.
    E) glucose.
    Answer: B

12. Teeth are primarily composed of a bone-like substance called
    A) crown.
    B) enamel.
    C) cementum.
    D) gingivae.
    E) dentin.
    Answer: E

13. The dentin of the root is covered by
    A) incisors.
    B) cuspids.
    C) premolars.
    D) molars.
    E) All of the above.
    Answer: C

14. The teeth which are adapted to cutting into food are the
    A) incisors.
    B) cuspids.
    C) premolars.
    D) molars.
    E) All of the above.
    Answer: A

15. Which cranial nerves control the secretion of saliva?
    A) VII and VIII
    B) VIII and X
    C) VII and IX
    D) IX and X
    E) X and XI
    Answer: C

16. All of the following are areas of the stomach EXCEPT
    A) cardia.
    B) duodenum.
    C) fundus.
    D) body.
    E) pylorus.
    Answer: B

17. Pepsinogen is produced by the
    A) hepatic cells in the liver.
    B) chief cells of the stomach.
    C) parietal cells of the stomach.
    D) exocrine glands of the pancreas.
    E) cells of the duodenum.
    Answer: C

18. The mucosa of the stomach, when empty, lies in large folds called
    A) gastric pits.
    B) gastric glands.
    C) rugae.
    D) peritoneum.
    E) omentum.
    Answer: C

19. The digestion of proteins by peptides starts in the
    A) oral cavity.
    B) esophagus.
    C) stomach.
    D) duodenum.
    E) ileum.
    Answer: C

20. The secretion of gastric juice and contraction of smooth muscle in the stomach wall
    are regulated by parasympathetic impulses transmitted via the
    A) hypoglossal nerves.
    B) spinal nerves.
    C) glossopharyngeal nerves.
    D) vagus nerves.
    E) facial nerves.
    Answer: D

21. The hormone that inhibits gastric emptying is
    A) gastric inhibitory protein.
    B) insulin.
    C) secretin.
    D) cholecytokinin.
    E) renin.
    Answer: D

22. Chyme is released by the stomach into the
    A) duodenum.
    B) esophagus.
    C) jejunum.
    D) ileum.
    E) cecum.
    Answer: A

23. The center for vomiting is located in the
    A) pons.
    B) cerebellum.
    C) medulla oblongata.
    D) hypothalamus.
    E) midbrain.
    Answer: C

24. Which of the following substances can be absorbed by the stomach?
    A) amino acids
    B) nucleic acids
    C) glucose
    D) fatty acids
    E) alcohol
    Answer: E

25. The G cells of the stomach secrete
    A) pepsinogen.
    B) pepsin.
    C) HCl.
    D) gastrin.
    E) CCK.
    Answer: D

26. The pH of the stomach is
    A) pH 9.
    B) pH 7.
    C) pH 5.
    D) pH 4.
    E) pH 2.
    Answer: E

27. The feeling of satiety is caused by
    A) cholecystokinin.
    B) secretin.
    C) pepsin.
    D) gastrin.
    E) All of the above.
    Answer: A

28. The pancreatic duct transports secretions from the pancreas to the
    A) stomach.
    B) liver.
    C) duodenum.
    D) jejunum.
    E) cecum.
    Answer: C

29. All of the following are substances found in pancreatic juice EXCEPT
    A) pepsin.
    B) trypsin.
    C) chymotrypsin.
    D) ribonuclease.
    E) lipase.
    Answer: A

30. Which of the following is NOT a protein-digesting enzyme?
    A) trypsin
    B) chymotrypsin
    C) carboxypeptidase
    D) amylase
    E) elastase
    Answer: D

31. The structure that divides the liver into two principal lobes is the
    A) greater omentum.
    B) falciform ligament.
    C) common hepatic duct.
    D) mesentery.
    E) pancreatic duct.
    Answer: B

32. Bile is produced by
    A) lymphocytes.
    B) Kupffer's cells.
    C) parietal cells.
    D) hepatocytes.
    E) chief cells.
    Answer: D

33. Substances absorbed by the small intestine are carried in the liver by the
    A) hepatic artery.
    B) hepatic vein.
    C) hepatic portal vein.
    D) common hepatic duct.
    E) central vein.
    Answer: C

34. All of the following are functions of the liver EXCEPT
    A) red blood cell production.
    B) storage of vitamins.
    C) synthesis of bile salts.
    D) excretion of bile.
    E) activation of Vitamin D.
    Answer: A

35. Most digestion and absorption occur in the
    A) oral cavity.
    B) esophagus.
    C) stomach.
    D) small intestine.
    E) large intestine.
    Answer: D

36. The final portion of the small intestine is the
    A) ileum.
    B) duodenum.
    C) cecum.
    D) jejunum.
    E) colon.
    Answer: A

37. Which of the following is NOT an enzyme in the breakdown of carbohydrates?
    A) pancreatic amylase
    B) salivary amylase
    C) maltase
    D) peptidase
    E) lactase
    Answer: D

38. Absorption in the small intestine occurs by
    A) diffusion.
    B) osmosis.
    C) facilitated diffusion.
    D) active transport.
    E) All of the above.
    Answer: E

39. Which of the following are absorbed into lacteals?
    A) monosaccharides
    B) amino acids
    C) triglycerides
    D) nucleic acids
    E) all of the above
    Answer: C

40. The first portion of the large intestine is the
    A) colon.
    B) cecum.
    C) anal canal.
    D) rectum.
    E) ileum.
    Answer: B

41. The appendix is attached to the
    A) cecum.
    B) ileum.
    C) ascending colon.
    D) descending colon.
    E) rectum.
    Answer: A

42. The large intestine absorbs
    A) water.
    B) electrolytes.
    C) vitamin B.
    D) vitamin K.
    E) All of the above.
    Answer: E

43. The portion of the large intestine just before the rectum is the
    A) anus.
    B) cecum.
    C) ascending colon.
    D) transverse colon.
    E) sigmoid colon.
    Answer: E

44. The feeding and satiety centers are located in the
    A) medulla oblongata.
    B) brain stem.
    C) hypothalamus.
    D) hippocampus.
    E) Both A and C.
    Answer: C

45. All of the following hormones increase appetite EXCEPT
    A) leptin.
    B) growth hormone releasing hormone.
    C) glucocorticoids.
    D) insulin.
    E) progesterone.
    Answer: A


## True-False

*Write T if the statement is true and F if the statement is false.*

1. Excretion is the elimination of indigestible substances from the gastrointestinal tract.
   Answer: False

2. The inner lining of the GI tract is a mucous membrane.
   Answer: True

3. The greater omentum contains many lymph nodes.
   Answer: True

4. The tongue is part of the GI tract.
   Answer: False

5. The lingual tonsil lies at the base of the tongue.
   Answer: True

6. The submandibular glands lie in front of the sublingual glands.
   Answer: False

7. Sympathetic stimulation results in dryness of the mouth.
   Answer: True

8. The enzyme salivary amylase starts the breakdown of proteins in the mouth.
   Answer: False

9. The esophagus lies behind the trachea.
   Answer: True

10. The muscularis of the stomach has three layers.
    Answer: True

11. Parasympathetic stimulation inhibits digestion.
    Answer: False

12. The presence of chyme in the duodenum slows gastric emptying.
    Answer: True

13. The liver is the heaviest gland of the body.
    Answer: True

14. The gallbladder produces bile.
    Answer: False

15. Paneth cells are found in the large intestine.
    Answer: False

## Short Answer

*Write the word or phrase that best completes each statement or answers the question.*

1. The medical specialty that deals with the structure, function, diagnosis, and treatment of the stomach and intestines is called _____.
   Answer: gastroenterology

2. The physiological term for eating is _____.
   Answer: ingestion

3. The waves of muscular contraction in the GI tract are referred to as _____.
   Answer: peristalsis

4. The mouth opens into the oropharynx through an opening called the _____.
   Answer: fauces

5. The food in the mouth is shaped into a rounded mass called the _____.
   Answer: bolus

6. The dentin of the crown of a tooth is covered by _____.
   Answer: enamel

7. Swallowing is divided into tree stages: the voluntary stage, the pharyngeal stage, and the _____ stage.
   Answer: esophageal

8. The large, central portion of the stomach is called the _____.
   Answer: body

9. Hydrochloric acid of the stomach is produced by _____.
   Answer: parietal cells

10. The enteric nervous system is found in the _____.
    Answer: submucosa

11. The intestinal hormone, which decreases gastric secretion, is ____.
    Answer: secretin

12. The acini of the pancreas produce ____.
    Answer: pancreatic juice

13. The common hepatic duct joins the cystic duct to form the ____.
    Answer: common bile duct

14. The principal bile pigment is ____.
    Answer: bilirubin

15. The hormone responsible for the ejection of bile from the gallbladder is ____.
    Answer: cholecystokinin (CCK)

16. The mucosa of the small intestine forms finger-like structures called ____.
    Answer: villi

17. Each villus of the small intestine has a lymphatic vessel called a ____.
    Answer: lacteal

18. The movements of the small intestine are peristalsis and ____.
    Answer: segmentation

19. Protein digestion starts with pepsin and is completed by the enzyme ____.
    Answer: peptidases

20. Bile salts act as emulsifying agents forming tiny spheres called ____.
    Answer: micelles

21. The inactive form of trypsin is activated by an enzyme called _____.
    Answer: enterokinase

22. The puckered appearance of the colon is caused by a series of pouches called ____.
    Answer: haustra

23. The last stage of digestion is carried out by ____.
    Answer: bacteria

24. Kupffer's cells are found in the _____.
    Answer: liver

25. An inflammation of the liver is called _____.
    Answer: hepatitis


## Matching

*Choose the item from Column 2 that best matches each item in Column 1.*

1.  Column 1: Inner lining of the GI tract
    Column 2: mucosa

2. Column 1: Outermost layer of the GI tract
   Column 2: serosa

3. Column 1: Secrete the hormone gastrin
   Column 2: G cells

4. Column 1: Produce hydrochloric acid
   Column 2: parietal cells

5. Column 1: Secrete pepsinogen
   Column 2: chief cells

6. Column 1: Secrete lysozyme
   Column 2: paneth cells

7. Column 1: "Good" cholesterol
   Column 2: high-density lipoproteins

8. Column 1: "Bad" cholesterol
   Column 2: low-density lipoproteins

9. Column 1: A type of food poisoning caused by a bacterial toxin.
   Column 2: Botulism

10. Column 1: A viral disease
    Column 2: Hepatitis


## Essay

*Write your answer in the space provided or on a separate sheet of paper.*

1. Name and describe the five basic activities of the digestive system.
   Answer: 1) Ingestion- The take-in of food through the mouth
   2) Mixing and movement of food along the digestive tract
   through muscle movement.
   3) Digestion- The chemical and mechanical breakdown of food.
   4) Absorption- The uptake of digested food from the GI tract into the
   circulation.
   5) Defecation- The elimination of indigestible substances from the GI tract.

2. Name and briefly describe the location of the different salivary glands.
   Answer: 1) Parotid glands are located under and in front of the ears between the skin
   And the masseter muscle.
   2) The submandibular glands lie beneath the base of the tongue in the floor of
   the mouth.
   3) The sublingual glands lie in front of the submandibular glands.

3. Name the functions of the liver.
   Answer: 1. Carbohydrate metabolism
   2. Lipid metabolism
   3. Protein metabolism
   4. Removal of drugs and hormones
   5. Excretion of bile

6.  Synthesis of bile salts
7.  Storage
8.  Phagocytosis
9.  Activation of Vitamin D

4.  Briefly describe the absorption of nutrients in the small intestine.
    Answer: Monosaccharides, amino acids, and short-chain fatty acids are absorbed into blood
        capillaries. Long-chain fatty acids and monoglycerides are absorbed as part of
        micelles, re-synthesized to triglycerides, and transported as chylomicrons.
        Chylomicrons are taken up by the lacteals.

## CHAPTER 20    Nutrition and Metabolism

## Multiple-Choice

*Choose the one alternative that best completes the statement or answers the question.*

1.  Neurons related to food intake are located in the
    A)  cervical spinal cord.
    B)  medulla oblongata.
    C)  pons.
    D)  hypothalamus.
    E)  cerebellum.
    Answer: D

2.  Minerals constitute about __ % of the total body weight.
    A)  2
    B)  4
    C)  6
    D)  8
    E)  10
    Answer: B

3.  Each gram of dietary protein provides about 4 Calories, a gram of fat provides about ___ Calories.
    A)  5
    B)  7
    C)  9
    D)  11
    E)  13
    Answer: C

4.  The vitamin essential to the absorption and utilization of calcium and phosphorus is vitamin ____.
    A)  C
    B)  E
    C)  A
    D)  B
    E)  D
    Answer: E

5.  The vitamin essential for the formation of photopigments of the retina is vitamin ____.
    A)  C
    B)  E
    C)  A
    D)  B
    E)  D
    Answer: C

6.  The term metabolism refers to
    A)  anabolic reactions.
    B)  catabolic reactions.
    C)  oxidation.
    D)  reduction.
    E)  All chemical reactions of the body.
    Answer: E

7. Which of the following depresses food intake?
   A) a warm environment
   B) low blood glucose levels
   C) decrease in adipose tissue
   D) a cold environment
   E) both A and B
   Answer: A

8. The breakdown of complex organic molecules into smaller ones is known as
   A) anabolism.
   B) catabolism.
   C) metabolism.
   D) oxidation.
   E) reduction.
   Answer: B

9. A chemical reaction that requires energy is
   A) anabolism.
   B) catabolism.
   C) metabolism.
   D) oxidation.
   E) All of the above.
   Answer: A

10. Chemically, enzymes are
    A) carbohydrates.
    B) proteins.
    C) lipids.
    D) nucleic acids.
    E) simple sugars.
    Answer: B

11. Oxidation refers to the
    A) breakdown of large molecules into smaller ones.
    B) synthesis of large molecules from smaller ones.
    C) the removal of water from a molecule.
    D) the removal of electrons and hydrogen ions from a molecule.
    E) the addition of electrons and hydrogen ions to a molecule.
    Answer: D

12. Galactose is converted to glucose by the
    A) stomach.
    B) bile.
    C) liver.
    D) blood.
    E) spleen.
    Answer: C

13. Each gram of a carbohydrate produces about
    A) 2.0 kilocalories.
    B) 4.0 kilocalories.
    C) 6.0 kilocalories.
    D) 8.0 kilocalories.
    E) 10. 0 kilocalories.
    Answer: B

14. Glucose is stored in the liver as
    A) starch.
    B) cellulose.
    C) fat.
    D) glycogen.
    E) ATP.
    Answer: D

15. The rate of glucose transport across cell membranes is greatly increased by
    A) insulin.
    B) glucagon.
    C) thyroxin.
    D) ADH.
    E) calcium.
    Answer: A

16. The net gain of ATP for each molecule of glucose which undergoes glycolysis is
    A) 10.
    B) 8.
    C) 6.
    D) 4.
    E) 2.
    Answer: E

17. Where in a cell does the Krebs cycle occur?
    A) nucleus
    B) cytoplasm
    C) mitochondria
    D) ribosomes
    E) Golgi apparatus
    Answer: C

18. Before entering the Krebs cycle, pyruvic acid has to be changed and combined with
    A) ATP.
    B) CoA.
    C) NADH.
    D) FADH.
    E) GTP.
    Answer: B

19. The largest amount of ATP molecules in cellular respiration are produced during the
    A) Krebs cycle.
    B) electron transport chain.
    C) glycolysis.
    D) transition reaction.
    E) fermentation.
    Answer: B

20. Glycogenolysis occurs when
    A) blood levels of glucose drop.
    B) glucagon is released.
    C) blood levels of glucose are high.
    D) Both A and B.
    E) Both B and C.
    Answer: D

21. The formation of glucose from proteins and fats is called
    A) glycogenolysis.
    B) lipolysis.
    C) glycogenesis.
    D) gluconeogenesis.
    E) lipogenesis.
    Answer: D

22. Which of the following hormones stimulate gluconeogenesis?
    A) thyroxine
    B) glucagon
    C) cortisol
    D) All of the above
    E) None of the above
    Answer: D

23. Ketone bodies are the result of catabolism of
    A) glycerol.
    B) fatty acids.
    C) amino acids.
    D) glycogen.
    E) glucose.
    Answer: B

24. During beta oxidation, fatty acids are converted into molecules of
    A) glucose.
    B) glyceraldehyde-3 phosphate.
    C) pyruvic acid.
    D) acetyl CoA.
    E) fructose-6-phosphate.
    Answer: D

25. Excess carbohydrates are synthesized into
    A) amino acids.
    B) proteins.
    C) triglycerides.
    D) ketone bodies.
    E) None of the above.
    Answer: C

26. During digestion, proteins are broken down into molecules of
    A) glucose.
    B) fatty acids.
    C) glycerol.
    D) amino acids.
    E) nucleic acids.
    Answer: D

27. Protein synthesis occurs in
    A) ribosomes.
    B) mitochondria.
    C) lysosomes.
    D) Golgi complexes.
    E) smooth endoplasmic reticulum.
    Answer: A

28. Which of the following minerals is considered a trace element?
    A) potassium
    B) magnesium
    C) sulfur
    D) chlorine
    E) iodine
    Answer: E

29. Which of the following vitamins can be made by the body?
    A) Vitamin C
    B) Vitamin D
    C) Vitamin E
    D) Vitamin K
    E) Vitamin B
    Answer: B

30. Which of the following vitamins is water-soluble?
    A) Vitamin B
    B) Vitamin A
    C) Vitamin D
    D) Vitamin E
    E) Vitamin K
    Answer: A

31. Which of the following factors affect the metabolic rate?
    A) exercise
    B) age
    C) gender
    D) hormones
    E) all of the above
    Answer: E

32. Which of the following hormones regulates the basal state of metabolism?
    A) vasopressin
    B) thyroxin
    C) insulin
    D) estrogen
    E) progesterone
    Answer: B

33. The transfer of heat from a warmer object to a cooler one without physical contact is called
    A) conduction.
    B) evaporation.
    C) convection.
    D) radiation.
    E) transformation.
    Answer: D

34. The thermostat of the human body is located in the
    A) thalamus.
    B) hypothalamus.
    C) cortex.
    D) medulla oblongata.
    E) limbic system.
    Answer: B

35. All of the following result in an increase of body temperature EXCEPT
    A) vasoconstriction.
    B) vasodilation.
    C) epinephrine release.
    D) release of thyroid hormones.
    E) shivering.
    Answer: B

36. In degrees of Celsius normal body temperature is near
    A) 35.
    B) 37.
    C) 39.
    D) 40.
    E) 41.
    Answer: B

37. Sympathetic stimulation
    A) increases cellular metabolism.
    B) decreases cellular metabolism.
    C) has no effect on cellular metabolism.
    D) decreases body temperature.
    E) cause vasodilation.
    Answer: A

38. Aspirin reduces fever by
    A) inhibiting the hypothalamus.
    B) inhibiting prostaglandin secretion.
    C) stimulating epinephrine release.
    D) decreasing metabolism.
    E) lowering thyroid function.
    Answer: B

39. Which of the following processes are part of the carbohydrate metabolisms?
    A) glycolysis
    B) electron transport chain
    C) the Krebs cycle
    D) formation of acetyl coenzyme A
    E) all of the above
    Answer: E

40. Coenzyme A is a derivative of
    A) vitamin D.
    B) vitamin E.
    C) vitamin B.
    D) vitamin C.
    E) None of the above.
    Answer: C

41. How many molecules of ATP are generated during the complete oxidation of a glucose molecule?
    A) 30 to 32
    B) 32 to 34
    C) 34 to 36
    D) 36 to 38
    E) 38 to 40
    Answer: D

42. Which of the following compounds transport dietary lipids to adipose tissue for storage?
    A) chylomicrons
    B) low density lipoproteins
    C) high density lipoproteins
    D) peripheral proteins
    E) phospholipids
    Answer: A

43. Which of the following compounds has the highest protein content?
    A) triglycerides
    B) very-low-density-lipoproteins
    C) low-density lipoproteins
    D) high-density lipoproteins
    E) chylomicrons
    Answer: D

44. Fatty acid metabolism begins in the
    A) cytoplasm.
    B) mitochondria.
    C) nucleus.
    D) smooth ER.
    E) ribosomes.
    Answer: B

45. Which of the following hormone(s) is/are the MAIN regulators of the basal metabolic rate?
    A) human growth hormone
    B) testosterone
    C) thyroid hormones
    D) insulin
    E) vasopressin
    Answer: C

## True-False

*Write T if the statement is true and F if the statement is false.*

1. Water is an excellent solvent and suspending medium.
   Answer: True

2. The hormone cholecytokinin is secreted when proteins enter the stomach.
   Answer: False

3. The milk, yogurt, and cheese group form the base of the food pyramid.
   Answer: False

4. In anabolism, simple substances combine into more complex molecules.
   Answer: True

5. Many coenzymes are derivatives of vitamins.
   Answer: True

6. The removal of electrons from a molecule is referred to as reduction.
   Answer: False

7. Oxidation is usually an energy-releasing reaction.
   Answer: True

8. The Krebs cycle is also called anaerobic cellular respiration.
   Answer: False

9. Glycogenesis decreases blood sugar levels.
   Answer: True

10. Approximately 80 percent of stored triglycerides are deposited in subcutaneous tissue.
    Answer: False

11. Hepatocytes synthesize lipids from glucose.
    Answer: True

12. Essential amino acids can be synthesized by body cells.
    Answer: False

13. Large quantities of vitamins are necessary to maintain normal metabolism.
    Answer: False

14. A kilocalorie is equal to 1000 calories
    Answer: True

15. The ingestion of food can raise the metabolic rate.
    Answer: True

## Short Answer

*Write the word of phrase that best completes each statement or answers the question.*

1. Organic nutrients that are required in small amounts to maintain normal metabolism are called _____.
   Answer: vitamins

2. Vitamins that can inactivate oxygen free radicals are called _____.
   Answer: antioxidants

3. The principal anion in extracellular fluid is ____.
   Answer: Cl⁻

4. The vitamin produced by intestinal bacteria and essential in the production of prothrombin is vitamin ___.
   Answer: K

5. The formation of peptide bonds between amino acids is a(n) ____ process.
   Answer: anabolic

6. The removal of electrons in the conversion of lactic acid into pyruvic acid is an example of ____.
   Answer: oxidation

7. The oxidation of glucose into two molecules of pyruvic acid is called ____.
   Answer: glycolysis

8. The food guide pyramid suggests ___ to ___ servings of the vegetable group per day.
   Answer: 3 to 5

9. The series of oxidation-reduction reactions that take place on the inner membrane of the mitochondria, And generate a large number of ATP molecules, is known as ____.
   Answer: electron transport chain

10. The formation of glycogen from glucose is called ____.
    Answer: glycogenesis

11. The splitting of triglyceride molecules is a process called ____.
    Answer: lipolysis

12. Before fatty acids can be metabolized by cells they have to undergo a series of reactions in the liver called ____.
    Answer: beta-oxidation

13. The removal of nitrogen from amino acids is called ____.
    Answer: deamination

14. Amino acids that can be synthesized by body cells are referred to as ____.
    Answer: nonessential amino acids

15. Vitamins that are stored in hepatocytes are ____ -soluble.
    Answer: fat

16. The higher the body temperature, the ____ the metabolic rate.
    Answer: higher

17. The metabolic rate ____ with age.
    Answer: decreases

18. The transfer of heat by movement of a liquid or gas between areas of different temperature is called ____.
    Answer: convection

19. The neurons of the hypothalamus involved in the regulation of body temperature are located in the ____.
    Answer: preoptic area

20. The transfer of body heat to a substance or object in contact with the body is called ____.
    Answer: conduction

21. Thyroid hormones will ____ body temperature.
    Answer: increase

22. The phase of fever when the skin becomes warm and the person begins to sweat is called the ____.
    Answer: crisis

23. The abbreviation BMR means ____.
    Answer: basal metabolic rate

24. The condition when body weight is 10 to 20 percent above the desirable standard due to fat accumulation is called ____.
    Answer: obesity

25. ____ refers to chemical reactions in the body.
    Answer: metabolism

## Matching

*Choose the item from Column 2 that best matches each item in Column 1.*

1. Column 1: Chemical reactions that require energy.
   Column 2: anabolic reactions

2. Column 1: Chemical reactions that release energy.
   Column 2: catabolic reactions

3. Column 1: The removal of electrons from a molecule.
   Column 2: oxidation

4. Column 1: The addition of electrons to a molecule.
   Column 2: reduction

5. Column 1: The formation of glucose from non-carbohydrates.
   Column 2: gluconeogenesis

6. Column 1: The conversion of glycogen to glucose.
   Column 2: glycogenolysis

7. Column 1: The formation of glycogen from glucose.
   Column 2: glycogenesis

8. Column 1: A water-soluble vitamin.
   Column 2: Vitamin C

9. Column 1: A fat-soluble vitamin
   Column 2: Vitamin A

## Essay

*Write your answer in the space provided or on a separate sheet of paper.*

1. Define nutrients and name the principal classes of nutrients.
   Answer: Nutrients are chemical substances in food that provide energy, or assist in the functioning of various body processes. The principal classes of nutrients are carbohydrates, lipids, proteins, minerals, vitamins, and water.

2. Name the guidelines for healthy eating.
   Answer:  1) Eat a variety of foods
       2)   Maintain healthy weight
       3)   Choose foods low in fat, saturated fat, and cholesterol
       4)   Eat plenty of vegetables, fruits, and grain products
       5)   Use sugars only in moderation
       6)   Use salt and sodium only in moderation
       7)   If you drink alcoholic beverages, do so in moderation

3. Name the steps of cellular respiration
   Answer: glycolysis, the Krebs cycle, and the electron transport chain

4. Name the principal routes by which the body can loose heat.
   Answer: radiation, evaporation, convection, and conduction

## CHAPTER 21    The Urinary System

## Multiple-Choice

*Choose the one alternative that best completes the statement or answers the question.*

1.  The kidneys
    A) help regulate blood volume.
    B) help to control blood pressure.
    C) secrete erythropoietin.
    D) help control blood pH.
    E) All of the above are correct.
    Answer: E

2.  The location of the kidneys in relationship to the peritoneal lining of the abdominal cavity is referred to
    as
    A) retroperitoneal.
    B) retroabdominal.
    C) dorsal.
    D) posterior.
    E) inferior.
    Answer: A

3.  All of the following belong to the urinary system EXCEPT
    A) urethra.
    B) ureter.
    C) bladder.
    D) prostate.
    E) kidneys.
    Answer: D

4.  Together with the skin and the liver, the kidney synthesize _____, the active form of vitamin D.
    A) calcitonin
    B) calcitriol
    C) erythropoietin
    D) renin
    E) angiotensin
    Answer: B

5.  The ureter leaves the kidney at the
    A) renal hilus.
    B) renal pelvis.
    C) renal capsule.
    D) renal pyramid.
    E) renal papillae.
    Answer: A

6.  The renal pyramids are located in the
    A) renal cortex.
    B) renal medulla.
    C) renal capsule.
    D) pelvis.
    E) papillae.
    Answer: B

7. The functional units of the kidneys are called
   A) glomeruli.
   B) calyces.
   C) nephrons.
   D) corpuscles.
   E) tubules.
   Answer: C

8. The blood vessel leading away from the glomerulus is the
   A) renal vein.
   B) renal artery.
   C) peritubular capillary.
   D) efferent arteriole.
   E) afferent arteriole.
   Answer: D

9. The Bowman's capsule is part of the
   A) renal pyramid.
   B) renal corpuscle.
   C) renal tubule.
   D) renal pelvis.
   E) minor calyx.
   Answer: B

10. The first portion of the renal tubule is the
    A) collecting duct.
    B) papillary duct.
    C) proximal convoluted tubule.
    D) distal convoluted tubule.
    E) loop of Henle.
    Answer: C

11. Which of the following helps to regulate renal blood pressure?
    A) renal tubule
    B) loop of Henle
    C) renal artery
    D) juxtaglomerular apparatus
    E) Bowman's Capsule
    Answer: D

12. The distal convoluted tubules empty into a
    A) minor calyx.
    B) loop of Henle.
    C) papillary duct.
    D) pelvis.
    E) collecting tubule.
    Answer: E

13. The interlobar arteries divide into the
    A) interlobular arteries.
    B) afferent arterioles.
    C) segmental arteries.
    D) arcuate arteries.
    E) peritubular capillaries.
    Answer: D

14. The peritubular capillaries drain into the
    A) arcuate veins.
    B) interlobular veins.
    C) interlobar veins.
    D) segmental veins.
    E) renal vein.
    Answer: B

15. The conversion of angiotensin I into angiotensin II occurs in the
    A) liver.
    B) pancreas.
    C) kidney.
    D) heart.
    E) lung.
    Answer: E

16. Which of the following represents the correct order of anatomical structures in the nephron?
    A) glomerulus, proximal convoluted tubules, loop of Henle, distal convoluted tubules
    B) proximal convoluted tubules, loop of Henle, distal convoluted tubules, glomerulus
    C) glomerulus, distal convoluted tubules, loop of Henle, proximal convoluted tubules
    D) distal convoluted tubules, glomerulus, proximal convoluted tubules, loop of Henle
    E) glomerulus, loop of Henle, proximal convoluted tubules, distal convoluted tubules
    Answer: A

17. The substance secreted by the juxtaglomerular cells is
    A) angiotensin I.
    B) angiotensin II.
    C) renin.
    D) ADH.
    E) angiotensionogen.
    Answer: C

18. The cells of the inner wall of the glomerulus are called
    A) macula densa.
    B) podocytes.
    C) glomerular cells.
    D) capsular cells.
    E) epithelial cells.
    Answer: B

19. On average ___ to ___ liters of urine are excreted per day.
    A) 1-2
    B) 2-3
    C) 3-4
    D) 4-5
    E) Less than all of the above.
    Answer: A

20. Which of the following is/are involved in the regulation of the GFR?
    A) myogenic mechanism
    B) tubuloglomerular feedback
    C) atrial natriuretic peptide
    D) Both A and B are involved.
    E) All of the above are involved.
    Answer: E

21. Most glucose molecules are reabsorbed in the
    A) proximal convoluted tubules.
    B) distal convoluted tubules.
    C) ascending limb of the loop of Henle.
    D) descending limb of the loop of Henle.
    E) collecting ducts.
    Answer: A

22. The reabsorption of chlorine ions is influenced by the movements of
    A) water.
    B) sodium ions.
    C) calcium ions.
    D) glucose molecules.
    E) potassium ions.
    Answer: B

23. The first step in urine formation is
    A) tubular secretion.
    B) secretion of ADH.
    C) tubular reabsorption.
    D) water reabsorption.
    E) glomerular filtration.
    Answer: E

24. Which of the following is found in plasma but not usually in glomerular filtrate?
    A) nitrogenous wastes
    B) large proteins
    C) vitamins
    D) water
    E) amino acids
    Answer: B

25. The glomerular filtration rate in a normal adult male is about
    A) 250 liters a day.
    B) 210 liters a day.
    C) 180 liters a day.
    D) 120 liters a day.
    E) 80 liters a day.
    Answer: C

26. Angiotensin II cause all of the following EXCEPT
    A) vasoconstriction of arterioles.
    B) stimulation of diuresis.
    C) stimulation of the thirst center.
    D) stimulation of ADH secretion.
    E) stimulation of aldosterone secretion.
    Answer: B

27. All of the following processes are involved in tubular reabsorption EXCEPT
    A) osmosis.
    B) diffusion.
    C) filtration.
    D) active transport.
    E) All of the above are part of tubular reabsorption.
    Answer: C

28. Which of the following substances can be eliminated from the blood by tubular secretions?
    A) potassium ions
    B) hydrogen ions
    C) ammonium ions
    D) penicillin
    E) All of the above.
    Answer: E

29. Which of the following diuretics inhibit ADH secretion?
    A) coffee
    B) tea
    C) beer
    D) wine
    E) Both C and D.
    Answer: E

30. From the collecting ducts urine drains into the
    A) renal pelvis.
    B) urinary bladder.
    C) minor calyces.
    D) major calyces.
    E) ureter.
    Answer: C

31. The structure that connects a kidney to the urinary bladder is the
    A) ureter.
    B) urethra.
    C) renal pelvis.
    D) collecting duct.
    E) calyx.
    Answer: A

32. Which of the following transport urine from the kidneys to the urinary bladder?
    A) peristalsis
    B) hydrostatic pressure
    C) gravity
    D) All of the above.
    E) None of the above.
    Answer: D

33. The urinary bladder is lined by
    A) pseudostratified epithelium.
    B) transitional epithelium.
    C) sqamous epithelium.
    D) cuboidal epithelium.
    E) columnar epithelium.
    Answer: B

34. Urine is expelled from the urinary bladder by
    A) excretion.
    B) defecation.
    C) micturition.
    D) filtration.
    E) None of the above.
    Answer: C

35. The terminal portion of the urinary system is the
    A) urinary bladder.
    B) calyx.
    C) ureter.
    D) urethra.
    E) collecting duct.
    Answer: D

36. Which of the following can decrease urine flow?
    A) nervousness.
    B) high temperatures.
    C) diuretics.
    D) increase water intake.
    E) inhibition of ADH.
    Answer: B

37. All of the following are principle solutes of urine EXCEPT
    A) urea.
    B) creatinine.
    C) glycogen.
    D) uric acid.
    E) urobilinogen.
    Answer: C

38. Bedwetting is referred to as
    A) enuresis.
    B) anuria.
    C) diuresis.
    D) dysuria.
    E) azotemia.
    Answer: A

39. Which of the following is an excretory organ?
    A) skin.
    B) lung.
    C) GI tract.
    D) All of the above.
    E) None of the above.
    Answer: D

40. The real arteries enter the kidneys at the
    A) minor calyx.
    B) major calyx.
    C) renal pelvis.
    D) renal hilus.
    E) renal pyramid.
    Answer: D

41. Osmoreceptors measuring the water concentration of the blood are located in the
    A) medulla oblongata.
    B) hypothalamus.
    C) urinary bladder.
    D) juxtaglomerular complex.
    E) adrenal medulla.
    Answer: B

42. Which of the following hormones increase reabsorption of water in the collecting tubules?
   A) renin
   B) ADH
   C) aldosterone
   D) insulin
   E) angiotensin
   Answer: B

43. What is the percentage of water in normal urine?
   A) 95%
   B) 87%
   C) 80%
   D) 75%
   E) 50%
   Answer: A

44. The average capacity of the urinary bladder is _____ to _____ mL.
   A) 100-200
   B) 300-400
   C) 500-600
   D) 700-800
   E) 900-1000
   Answer: D

45. On average the renal blood flow declines by ___ between the ages of 40 and 70.
   A) 20%
   B) 30%
   C) 40%
   D) 50%
   E) 60%
   Answer: D


## True-False

*Write T if the statement is true and F if the statement is false.*

1. The kidneys help regulate the blood pressure.
   Answer: True

2. The renal columns are part of the renal pyramids.
   Answer: False

3. The renal corpuscle filters plasma into the renal tubule.
   Answer: True

4. The proximal convoluted tubules are continuous with the ascending limb of Henle.
   Answer: False

5. The major work of the urinary system is done by the nephron.
   Answer: True

6. The blood pressure in the kidneys is about 56mm Hg.
   Answer: False

7. The glomerular filtration rate is primarily regulated by the nervous system.
   Answer: False

8. Water retention increases blood volume and therefore blood pressure.
   Answer: True

9. Hemodialysis is typically performed twice a week.
   Answer: False

10. The opening and closing of the internal urethral sphincter is involuntary.
    Answer: True

11. The urethra is the tube connecting a kidney with the urinary bladder.
    Answer: False

12. Urine volume is influenced by blood pressure.
    Answer: True

13. Constriction of the afferent arteriole decreases blood flow into the glomerulus, which increases GFR.
    Answer: False

14. The kidneys produce erythropoietin.
    Answer: True

15. Pyelitis is the inflammation of the urinary bladder.
    Answer: False

## Short Answer

*Write the word or phrase that best completes each statement or answers the question.*

1. A toxic level of urea in the blood is called ____.
   Answer: uremia

2. The kidneys are located outside the peritoneal lining of the abdominal cavity, therefore their position is described as ____.
   Answer: retroperitoneal

3. Each kidney is enclosed in a transparent, fibrous membrane called ____.
   Answer:  renal capsule

4. The large cavity close to the renal hilus is the ____.
   Answer: renal pelvis

5. Urine formation involves three processes: 1) glomerular filtration, 2) tubular reabsorption, and 3) ____.
   Answer:  tubular secretion

6. The ability of the kidneys to maintain a constant blood pressure despite changes in the systemic blood pressure is due to ____.
   Answer: renal autoregulation

7. Renal autoregulation involves the macula densa cells of the ____.
   Answer: juxtaglomerular apparatus

8. The hormone that promotes excretion of water and excretion of sodium is ____.
   Answer: atrial natriuretic peptide

9. The maximum amount of a substance that can be reabsorbed by the renal tubules is called ____.
   Answer: tubular maximum (renal threshold)

10. The presence of glucose in the urine is referred to as ____.
    Answer: glycosuria

11. The process of filtering blood by artificial means is called ____.
    Answer: hemodialysis

12. The small triangular area at the base of the urinary bladder is called ____.
    Answer: trigone

13. A lack of voluntary control over micturition is referred to as ____.
    Answer: incontinence

14. An analysis of the volume, physical and chemical properties of urine is called ____.
    Answer:  urinalysis

15. A progressive and generally irreversible decline in the glomerular filtration rate is called ____.
    Answer: chronic renal failure

16. An inflammation of the urinary bladder is called ____.
    Answer: cystitis

17. The functional unit of the kidney is a (n) ____.
    Answer: nephron

18. The hormone that increases water reabsorption is ____.
    Answer: ADH (antidiuretic hormone)

19. The tube that discharges urine from the body is called the ____.
    Answer: urethra

20. Excessive urine is referred to as ____.
    Answer: polyuria

21. The scientific study of the kidney is called ____.
    Answer: nephrology

22. Urine formed by the nephrons drains into collecting ducts and then into larger ____ ducts.
    Answer: papillary

23. The amount of filtrate produced by the kidneys per minute is referred to as ___ ___ ___.
    Answer: glomerular filtration rate

24. The glomerular filtrate is also called ___ ___.
    Answer: tubular fluid

25. A condition that is the result of inadequate release of ADH is ____ ____.
    Answer: diabetes insipidus

## Matching

*Choose the item from Column 2 that best matches each item in Column 1.*

1. Column 1: an inflammation of the kidney that involves glomeruli.
   Column 2: glomerulonephritis

2. Column 1: An inflammation of the renal pelvis and its calyces.
   Column 2: pyelitis

3. Column 1: A condition that refers to protein in the urine.
   Column 2: nephrotic syndrome

4. Column 1: A decrease or cessation of glomerular filtration.
   Column 2: renal failure

5. Column 1: Presence of urea or other nitrogenous elements in the blood.
   Column 2: azotemia

6. Column 1: Hernia of the urinary bladder.
   Column 2: cystocele

7. Column 1: Increased excretion of urine.
   Column 2: diuresis

8. Column 1: Painful urination.
   Column 2: dysuria

9. Column 1: Toxic levels of urea in the blood.
   Column 2: uremia

10. Column 1: Narrowing of the lumen of the ureter or urethra.
    Column 2: stricture

## Essay

*Write your answer in the space provided or on a separate sheet of paper.*

1. Describe the functions of the kidneys.
   Answer: 1. They regulate blood volume and composition of the blood; they remove waste and excrete selected amounts of wastes to include hydrogen ions, which helps control the pH of blood.
   2. They help regulate blood pressure by secreting renin, an enzyme which activates the renin-angiotensin system.
   3. They contribute to metabolism by 1) performing gluconeogenesis during starvation, 2) secreting erythropoietin which stimulates red blood cell production, and 3) participating in calcitriol synthesis, the active form of Vitamin D.

2. Name the components of a nephron in correct order, starting with the glomerolus.
   Answer: glomerolus-Bowman's capsule-proximal convoluted tubules-descending limb of the Loop of Henle-ascending limb of the loop of Henle-distal convoluted tubules-collecting duct.

3. Name the actions of angiotensin II.
   Answer: 1. Vasoconstriction of arterioles
   2. Stimulation of aldosterone secretion by the adrenal cortex

3. Stimulation of the thirst center in the hypothalamus
4. Stimulation of ADH secretion from the posterior pituitary

4. Define glomerular filtration rate.
   Answer: The glomerular filtration rate (GFR) is the amount of filtrate formed in both kidneys every minute.

## CHAPTER 22   Fluid, Electrolyte, and Acid-Base Balance

## Multiple-Choice

*Choose the one alternative that best completes the statement or answers the question.*

1.  The body fluids found within the cells is called
    A)  plasma.
    B)  extracellular fluid.
    C)  interstitial fluid.
    D)  intracellular fluid.
    E)  water.
    Answer: D

2.  In lean adults body fluid makes about ___-___% of total body mass.
    A)  30-40
    B)  45-50
    C)  55-60
    D)  65-70
    E)  85-90
    Answer: C

3.  Blood plasma belongs to
    A)  intracellular fluid.
    B)  extracellular fluid.
    C)  interstitial fluid.
    D)  Both B and C.
    E)  None of the above.
    Answer: B

4.  Peritoneal fluids between serous membranes belongs to
    A)  intracellular fluid.
    B)  extracellular fluid.
    C)  interstitial fluid.
    D)  Both B and C.
    E)  None of the above.
    Answer: B

5.  Which of the following extracellular fluids is found in the ear?
    A)  aqueous humor
    B)  endolymph
    C)  peritoneal fluid
    D)  synovial fluid
    E)  cerebrospinal fluid
    Answer: B

6.  Synovial fluid is found in
    A)  the inner ear.
    B)  the middle ear.
    C)  joints.
    D)  serous membranes.
    E)  Both A and B.
    Answer: C

7. Most solutes in body fluid are
   A) electrolytes.
   B) proteins.
   C) sugars.
   D) amino acids.
   E) lipids.
   Answer: A

8. The highest proportion of water in body weight is found in
   A) lean women.
   B) lean men.
   C) fat women.
   D) fat men.
   E) infants.
   Answer: E

9. Water lost through the actions of the
   A) kidneys.
   B) gastrointestinal tract.
   C) lungs.
   D) skin.
   E) All of the above.
   Answer: E

10. The center for thirst is located in the
    A) medulla oblongata
    B) hypothalamus
    C) pituitary
    D) kidney
    E) GI tract
    Answer: B

11. Dehydration stimulates the center for thirst by
    A) decreased production of saliva.
    B) increased osmotic pressure.
    C) decreased blood volume.
    D) B and C are correct.
    E) A, B, and C are correct.
    Answer: E

12. Total fluid output of the body under normal conditions is approximately
    A) 200 ml/day.
    B) 2500 ml/day.
    C) 400 ml/day.
    D) 1500 ml/day.
    E) 4300 ml/day.
    Answer: E

13. All of the following are a result of dehydration EXCEPT
    A) high osmotic pressure of the blood.
    B) decreased production of saliva.
    C) decreased production of blood.
    D) decreased blood volume.
    E) decreased blood pressure.
    Answer: C

14. All of the following are involved in regulation of **fluid output** EXCEPT
   A) ADH.
   B) ANP.
   C) CCK.
   D) Renin.
   E) Aldosterone.
   Answer: C

15. High blood pressure
   A) decreases urine output.
   B) increases the glomerular filtration rate.
   C) increases osmolarity of the blood.
   D) causes excessive water loss through the skin.
   E) causes increased fluid loss through the lungs.
   Answer: B

16. Glucose, urea, and creatine are considered to be
   A) electrolytes.
   B) nonelectrolytes.
   C) cations.
   D) solvents.
   E) inorganic compounds.
   Answer: B

17. Which of the following hormones is responsible for **increased urinary** excretion of $Na^+$?
   A) angiotensin II
   B) aldosterone
   C) antidiuretic hormone
   D) atrial natriuretic peptide
   E) epinephrine
   Answer: D

18. Release of ADH is stimulated by
   A) low osmotic pressure.
   B) alcohol consumption.
   C) increased blood volume.
   D) increased osmotic pressure of plasma.
   E) All of the above.
   Answer: D

19. All of the following are functions of electrolytes **EXCEPT**
   A) control osmosis between body compartments.
   B) are necessary for the production of action **potentials.**
   C) control bulk flow between plasma and inter**stitial fluid.**
   D) help maintain the acid base balance.
   E) control the secretion of some hormones.
   Answer: C

20. Osmotic pressure is related to
   A) the number of particles in a solution.
   B) the size of particles in a solution.
   C) the weight of particles in a solution.
   D) All of the above.
   E) None of the above.
   Answer: A

21. The major difference between plasma and interstitial fluid is
    A) the osmolarity.
    B) the concentration of carbohydrates.
    C) the concentration of solutes.
    D) the concentration of protein anions.
    E) the concentration of water.
    Answer: D

22. The most abundant cation in extracellular fluid is
    A) calcium.
    B) sodium.
    C) potassium.
    D) magnesium.
    E) hydrogen.
    Answer: B

23. The level of sodium in the blood is controlled by
    A) aldosterone.
    B) ADH.
    C) ANP.
    D) A and B.
    E) All of the above.
    Answer: E

24. ANP is produced in the
    A) kidneys.
    B) adrenal glands.
    C) pituitary gland.
    D) hypothalamus.
    E) atria of the heart.
    Answer: E

25. All of the following are characteristics of hyponatremia EXCEPT
    A) intense thirst.
    B) muscular weakness.
    C) dizziness.
    D) hypotension.
    E) tachycardia.
    Answer: A

26. A lower than normal level of potassium is referred to as
    A) hyperkalemia.
    B) hypokalemia.
    C) hyponatremia.
    D) hypernatrimia.
    E) hypochloremia.
    Answer: B

27. The levels of calcium in the blood are regulated by
    A) ADH and parathyroid hormone.
    B) aldosterone and ADH.
    C) parathyroid hormone and calcitonin.
    D) calcitonin and renin.
    E) renin and aldosterone.
    Answer: C

28. Hypercalcemia is characterized by
    A) lethargy.
    B) bone pain.
    C) muscle cramps.
    D) numbness and tingling of fingers.
    E) both A and B.
    Answer: E

29. The magnesium level is regulated by
    A) parathyroid hormone.
    B) antidiuretic hormone.
    C) aldosterone.
    D) angiotensin.
    E) adrenalin.
    Answer: C

30. Which of the following is the major extracellular anion?
    A) phosphate
    B) calcium
    C) magnesium
    D) chloride
    E) potassium
    Answer: D

31. Which of the following ions diffuses easily between extracellular and intracellular compartments?
    A) magnesium
    B) chlorine
    C) sodium
    D) phosphate
    E) potassium
    Answer: B

32. Which of the following is characteristic in renal insufficiency?
    A) hyperphosphatemia
    B) hypophosphatemia
    C) hyponatremia
    D) hypernatrimia
    E) hypocalcemia
    Answer: A

33. Which of the following is a function of ions in the body?
    A) help maintain the acid-base balance
    B) control of water movement between fluid compartments
    C) carry electrical impulses
    D) needed for enzyme activity
    E) All of the above.
    Answer: E

34. The concentration of ions is usually expressed in
    A) milliequivalents/liter
    B) units/liter
    C) units/milliliter
    D) milliequivalents/gram
    E) milliequivalents/100ml
    Answer: A

35. Under normal conditions, any fluid not reabsorbed by the venous end of the capillaries
    A) remains in the extracellular spaces.
    B) is absorbed by the surrounding cells.
    C) passes into lymphatic capillaries.
    D) is excreted in perspiration.
    E) diffuses to other capillary beds.
    Answer: C

36. The pH of the extracellular fluid of a healthy person is
    A) acidic.
    B) slightly acidic.
    C) neutral.
    D) slightly basic.
    E) basic.
    Answer: D

37. The acid-base balance depends on the concentration of
    A) calcium ions.
    B) hydrogen ions.
    C) potassium ions.
    D) sodium ions.
    E) magnesium ions.
    Answer: B

38. A substance that dissociates into hydroxyl ions is a(n)
    A) buffer.
    B) acid.
    C) base.
    D) salt.
    E) ion.
    Answer: C

39. Which of the following buffer systems help regulate the pH in red blood cells?
    A) carbonic acid buffer system
    B) phosphate buffer system
    C) hemoglobin buffer system
    D) protein buffer system
    E) bicarbonate buffer system
    Answer: B

40. The most abundant buffer system in body cells and plasma is the
    A) carbonic acid buffer system.
    B) bicarbonate buffer system.
    C) phosphate buffer system.
    D) hemoglobin buffer system.
    E) protein buffer system.
    Answer: E

41. The chemoreceptors in the medulla oblongata are stimulated by
    A) a decrease in the hydrogen ions of the blood.
    B) an increase in the hydrogen ions of the blood.
    C) decreased carbon dioxide concentration in the blood.
    D) decreased oxygen concentrations in the blood.
    E) an increase in the pH of blood.
    Answer: B

42. An increase in the carbon dioxide concentration in body fluids results in
    A) lower pH.
    B) more acid fluids.
    C) higher pH.
    D) A and B.
    E) B and C.
    Answer: D

43. Increased respiration
    A) removes carbon dioxide from the blood.
    B) reduces the hydrogen ion concentration in blood.
    C) increases the pH of blood.
    D) All of the above.
    E) None of the above.
    Answer: D

44. The principal physiological effect of acidosis is
    A) excitation of the central nervous system.
    B) depression of the central nervous system.
    C) excitation of cardiovascular center.
    D) overexcitabilty of peripheral nerves.
    E) inhibition of the neuromuscular junctions.
    Answer: B

45. The most abundant mineral in the body is
    A) magnesium.
    B) potassium.
    C) calcium.
    D) ion.
    E) sodium.
    Answer: C

# True-False

*Write T if the statement is true and F if the statement is false.*

1. Electrolytes are the main component of bodily fluids.
   Answer: False

2. Dehydration decreases the production of saliva.
   Answer: True

3. Excessive fluid in the blood results in hypotension.
   Answer: False

4. Glucose is one of the major electrolytes of the body.
   Answer: False

5. The ion concentration of extracellular fluid differs considerably from the ion concentration of intracellular fluids.
   Answer: True

6. A decrease in the sodium concentration causes the release of ADH from the posterior pituitary.
   Answer: False

7. Calcium is involved in blood clotting.
   Answer: True

8. Aldosterone is the main hormone in the homeostasis of calcium.
   Answer: False

9. Phosphate is mainly an intracellular electrolyte.
   Answer: True

10. PTH stimulates osteoclasts to release phosphate from mineral slats of bone matrix.
    Answer: True

11. Both body water and sodium are lost during excessive sweating.
    Answer: True

12. Lost body fluids should be replaced by drinking pure water.
    Answer: False

13. The majority of hydrogen ions are the result of cellular metabolism.
    Answer: True

14. Sodium is a strong acid and can be neutralized by bicarbonate.
    Answer: False

15. Respiration plays a role in the maintenance of blood pH.
    Answer: True

## Short Answer

*Write the word or phrase that best completes each statement or answers the question.*

1. Body fluids outside of cells are called ____.
   Answer: extracellular fluid

2. When water loss is greater than water gain the result is ____.
   Answer: dehydration

3. The thirst center is located in the ____.
   Answer: hypothalamus

4. The hormone that increases fluid loss in the urine is ____.
   Answer: atrial natriuretic peptide (ANP)

5. Negatively charged ions called ____.
   Answer: anions

6. Compounds that can conduct electric currents are called ____.
   Answer: electrolytes

7. In osmosis, water moves from an area of ____ osmotic pressure to an area with ____ osmotic pressure.
   Answer: lower, higher

8. The concentration of an ion in a solution is commonly expressed in ____.
   Answer: milliequivalents per liter

9.  A higher-than-normal blood sodium level is called ____.
    Answer: hypernatremia

10. The levels of blood calcium are regulated by the hormones ____ and ____.
    Answer: calcitonin, parathyroid hormone

11. The ion that activates enzymes for the carbohydrate and protein metabolism, triggers the sodium/potassium pump, and preserves the structure of DNA and RNA is ____.
    Answer: magnesium

12. The difference between the forces that move fluid out of the plasma and the forces that push it into the plasma is the ____.
    Answer: net filtration pressure

13. The principal cation inside the cell is ____.
    Answer: potassium

14. When carbon dioxide combines with water it forms ____.
    Answer: carbonic acid

15. A substance that ionizes to produce hydrogen ions is a (n) ____.
    Answer: acid

16. Systems that prevent drastic changes in the pH of bodily fluids are called ____.
    Answer: buffer systems (buffers)

17. The buffer system that buffers carbonic acid in the blood is the ____.
    Answer: hemoglobin buffer system

18. The inspiratory area is located in the ____.
    Answer: medulla oblongata

19. When blood pH decreases below 7.35, ____ occurs.
    Answer: acidosis

20. The physiological response to an acid-base imbalance to normalize blood pH is called ____.
    Answer: compensation

21. Plasma makes up ____% of the ECF.
    Answer: 20

22. The hormone responsible for decreased water loss in urine is ____.
    Answer: ADH

23. Substances that dissociate in water and form ions are ____.
    Answer: electrolytes

24. Renal compensation occurs in persons which have an altered blood pH due to ____ ____.
    Answer: respiratory causes

25. Water that is generated during the cell's chemical activity is called ____ water.
    Answer: metabolic

## Matching

*Choose the item from Column 2 that best matches each item in Column 1.*

1.  Column 1: muscular weakness, dizziness, headache, hypotension, tachycardia, and shock
    Column 2: hyponatremia

2.  Column 1: intense thirst, fatigue, restlessness, agitation, and coma
    Column 2: hypernatremia

3.  Column 1: cramps, fatigue, flaccid paralysis, nausea, vomiting, mental confusion, increased urine
    output, shallow respiration, and changes in the electrocardiogram
    Column 2: hypokalemia

4.  Column 1: irritability, anxiety, abdominal cramping, diarrhea, weakness, and paresthesia
    Column 2: hyperkalemia

5.  Column 1: numbness and tingling of the fingers, hyperactive reflexes, muscle cramps, tetany, and
    convulsions.
    Column 2: hypocalcemia

6.  Column 1: lethargy, weakness, anorexia, nausea, vomiting, polyuria, itching, bone pain, depression,
    confusion, paresthesia, stupor, and coma.
    Column 2: hypercalcemia

7.  Column 1: muscle spasms, alkalosis, depressed respiration, and coma
    Column 2: hypochloremia

## Essay

*Write your answer in the space provided or on a separate sheet of paper.*

1.  Name the major ions involved in the fluid and electrolyte balance and briefly describe their major
    location.
    Answer:  1.  Sodium is the most abundant extracellular ion and represents about 90 percent of
    extracellular cations.
    2.  Potassium is the most abundant cation in the intracellular fluid.
    3.  Calcium is the most abundant ion in the body and is principally an extracellular
    extracellular electrolyte.
    4.  Magnesium is primarily an intracellular electrolyte.
    5.  Chloride is the major extracellular anion.
    6.  Phosphate is principally an intracellular electrolyte.

2.  Explain the ways by which dehydration stimulates thirst.
    Answer: 1. Dehydration decreases the production of saliva, which causes dryness of the oral mucosa.
    A dry mouth and pharynx stimulates the osmoreceptors of the hypothalamus, resulting in
    the sensation of thirst.
    2.  Dehydration increases the osmotic pressure of blood, which also stimulates the
    osmoreceptors of the hypothalamus, resulting in thirst.
    3.  Dehydration decreases the blood volume, which causes a drop in blood pressure. This
    stimulates the renin-angiotensin II system. Angiotensin II stimulates the thirst center of
    the hypothalamus.

3. Explain water intoxication.

   Answer: If lost bodily fluids are replaced by pure water, body fluids become more dilute, which causes the concentration of sodium to fall below normal. A decrease in the sodium concentration causes the osmotic pressure of the interstitial fluid to fall. The result is osmosis of water into intracellular fluid. The cells become hypotonic and swell, a condition called water intoxication. Water intoxication causes disorientation, convulsions, coma, and possible death.

## CHAPTER 23   The Reproductive Systems

## Multiple-Choice

*Choose the one alternative that best completes the statement or answers the question.*

1.  Sperm production occurs in the
    A) Sertoli cells.
    B) seminiferous tubules.
    C) interstitial cells of Leydig.
    D) epididymis.
    E) vas deferens.
    Answer: B

2.  Which of the following cells is the most immature?
    A) sperm cell
    B) spermatid
    C) spermatogonia
    D) primary spermatocyte
    E) secondary spermatocyte
    Answer: C

3.  In the male, the hormone inhibin is produced in the
    A) hypothalamus.
    B) Sertoli cells.
    C) pituitary.
    D) interstitial cells of Leydig.
    E) spermatogonia.
    Answer: B

4.  The cell produced by fertilization is called
    A) gamete.
    B) secondary oocyte.
    C) embryo.
    D) zygote.
    E) fetus.
    Answer: D

5.  All of the following are true for meiosis EXCEPT
    A) it occurs during spermatogenesis.
    B) it occurs during oogenesis.
    C) the daughter cells have only half the chromosomes of the parent cell.
    D) it is subdivided into meiosis I and meiosis II.
    E) it can be used for cell repair.
    Answer: E

6.  During spermatogenesis, the cells formed by equatorial division are called
    A) spermatids.
    B) sperm cells.
    C) primary spermatocytes.
    D) secondary spermatocytes.
    E) spermatogonia.
    Answer: A

7.  The production of testosterone in the interstitial cells of Leydig is stimulated by
    A)  inhibin.
    B)  luteinizing hormone.
    C)  follicle-stimulating hormone.
    D)  relaxin.
    E)  progesterone.
    Answer: B

8.  The hormones that control the rate of spermatogenesis are
    A)  testosterone and inhibin.
    B)  testosterone and FSH.
    C)  inhibin and LH.
    D)  FSH and inhibin.
    E)  FSH and LH.
    Answer: D

9.  Sperm maturation occurs in the
    A)  seminiferous tubules.
    B)  straight tubules.
    C)  epididymis.
    D)  rete testes.
    E)  vas deferens.
    Answer: C

10. The ductus epididymis becomes less convoluted and is then referred to as
    A)  ductus deferens.
    B)  spermatic cord.
    C)  rete testes.
    D)  straight tubule.
    E)  ejaculatory duct.
    Answer: C

11. The terminal duct of the male reproductive system is the
    A)  ejaculatory duct.
    B)  penis.
    C)  urethra.
    D)  ductus deferens.
    E)  ureter.
    Answer:  C

12. 60 percent of the volume of semen is produced by the
    A)  prostate gland.
    B)  Cowper's gland.
    C)  seminal vesicles.
    D)  testis.
    E)  bulbourethral gland.
    Answer: C

13. The gland that surrounds the superior portion of the urethra in males is the
    A)  Cowper's gland.
    B)  prostate gland.
    C)  bulbourethral gland.
    D)  adrenal gland.
    E)  None of the above.
    Answer: B

14. Semen production requires the
    A) testis.
    B) prostate gland.
    C) bulbourethral gland.
    D) seminal vesicles.
    E) All of the above.
    Answer: E

15. The structure of the sperm cell containing mitochondria is the
    A) acrosome.
    B) head.
    C) tail.
    D) midpiece.
    E) center.
    Answer: D

16. All of the following are structures of the penis EXCEPT
    A) corpus spongiosum.
    B) prepuce.
    C) external urethral orifice.
    D) glands penis.
    E) ejaculatory duct.
    Answer: E

17. The average volume of semen for an ejaculation is
    A) 1 to 2 mL.
    B) 1.5 to 3mL.
    C) 2.5 to 5mL.
    D) 4.5 to 6mL.
    E) 5 to 7 mL.
    Answer: C

18. Infertility in males becomes likely if the sperm count falls below
    A) 50 million sperm/mL.
    B) 40 million sperm/mL.
    C) 30 million sperm/mL.
    D) 20 million sperm/mL.
    E) 10 million sperm/mL.
    Answer: D

19. The enzyme of sperm cells necessary to penetrate an ovum is located in the
    A) head of the sperm cell.
    B) acrosome.
    C) midpiece.
    D) tail portion of the sperm cell.
    E) nucleus of the sperm.
    Answer: B

20. All of the following are true for testosterone EXCEPT:
    A) it is an anabolic steroid.
    B) it stimulates the sex drive.
    C) it stimulates inhibin production.
    D) it is responsible for the secondary sex characteristics of males.
    E) All of the above are true.
    Answer: C

21. Spermatids contain _____ chromosomes.
    A) 16
    B) 20
    C) 23
    D) 36
    E) 46
    Answer: C

22. Sperm cells can survive up to ___ hours in the female reproductive tract.
    A) 10
    B) 24
    C) 36
    D) 48
    E) 52
    Answer: D

23. When follicular fluid builds up in the cavity of the follicles, it is called a(n)
    A) primary oocyte.
    B) polar body.
    C) primary follicle.
    D) secondary follicle.
    E) corpus luteum.
    Answer: D

24. The ovarian follicle that ruptures during ovulation is the
    A) primary follicle.
    B) secondary follicle.
    C) tertiary follicle.
    D) Graafian follicle.
    E) None of the above.
    Answer: D

25. At ovulation, a secondary oocyte is released into the
    A) uterus.
    B) uterine tubes.
    C) cortex of the ovary.
    D) cervix.
    E) infundibulum.
    Answer: E

26. Fertilization usually occurs in the
    A) vagina.
    B) cervix.
    C) uterus.
    D) uterine tubes.
    E) ovaries.
    Answer: D

27. The layer of the uterine wall that is shed during menstruation is the
    A) stratum functionalis.
    B) stratum basale.
    C) perimetrium.
    D) myometrium.
    E) Both A and B.
    Answer: A

28. The external genitalia of the **female are collectiv**ely called
    A) labia.
    B) vulva.
    C) clitoris.
    D) mons pubis.
    E) glans.
    Answer: B

29. The structure of the female **reproductive system that** contains erectile tissue is the
    A) eternal urethral orifice.
    B) vaginal orifice.
    C) labia minora.
    D) mons pubis.
    E) clitoris.
    Answer: E

30. All of the following are **structures of the mammary** glands EXCEPT
    A) Skene's gland.
    B) alveoli.
    C) Cooper's ligaments.
    D) areola.
    E) lobules.
    Answer: A

31. Which of the following **control the menstrual and** ovarian cycles of the female?
    A) GnRH
    B) FSH
    C) LH
    D) Estrogen
    E) All of the above.
    Answer: E

32. The initial secretion of estrogens by the growing ovarian follicles is directly stimulated by
    A) GnRH.
    B) relaxin.
    C) progesterone.
    D) FSH.
    E) LH.
    Answer: D

33. The hormone that works **with estrogen to prepare** the endometrium for implantation of a fertilized egg
    is
    A) LH.
    B) FSH.
    C) prolactin.
    D) progesterone.
    E) relaxin.
    Answer: D

34. The average menstrual cycle **is**
    A) 14 days.
    B) 18 days.
    C) 24 days.
    D) 28 days.
    E) 32 days.
    Answer: D

35. The rupture of the mature follicle during ovulation is due to a surge in
    A) FSH.
    B) LH.
    C) progesterone.
    D) estrogen.
    E) inhibin.
    Answer: B

36. The presence of which hormone in the maternal blood or urine indicates pregnancy
    A) hCG.
    B) FSH.
    C) LH.
    D) GnRH.
    E) progesterone.
    Answer: A

37. Currently the most prevalent sexually transmitted disease in the U.S. is
    A) AIDS.
    B) gonorrhea.
    C) chlamydia.
    D) genital herpes.
    E) syphilis.
    Answer: C

38. Which of the following sexually transmitted diseases is caused by the human papilomavirus?
    A) AIDS
    B) chlamydia
    C) genital herpes
    D) genital warts
    E) syphilis
    Answer: D

39. Which of the following is caused by bacteria?
    A) AIDS
    B) genital warts
    C) genital herpes
    D) trichomoniasis
    E) chlamydia
    Answer: E

40. All of the following can be cured EXCEPT
    A) genital herpes.
    B) chlamydia.
    C) syphilis.
    D) gonorrhea.
    E) trichomoniasis.
    Answer: A

41. The structure between the uterus and the vagina is the
    A) uterine tube.
    B) isthmus.
    C) cervix.
    D) vulva.
    E) hymen.
    Answer: C

42. "Crossing-over" occurs during
    A) spermatogenesis.
    B) oogenesis.
    C) meiosis I.
    D) meiosis II.
    E) mitosis.
    Answer: C

43. The circular pigmented area of skin surrounding the nipple of the mammary glands is called
    A) alveolus.
    B) areola.
    C) perineum.
    D) apex.
    E) vulva.
    Answer: B

44. The release of estrogen and inhibin by the dominant follicle decreases the release of
    A) progesterone
    B) GnRh
    C) LH
    D) FSH
    E) relaxin
    Answer: D

45. The absence of menstruation is referred to as
    A) hypospadias.
    B) PID.
    C) amenorrhea.
    D) dysmenorrhea.
    E) smegma.
    Answer: C


## True-False

*Write T if the statement is true and F if the statement is false.*

1. The production and survival of sperm require lower than normal body temperature.
   Answer: True

2. Sperm cells mature in the testis.
   Answer: False

3. Gametes differ from all other body cells because they contain the diploid chromosome number.
   Answer: False

4. Sperm are produced at the rate of about 300 million per day.
   Answer: True

5. FSH stimulates testosterone production.
   Answer: False

6. The onset of puberty in males is signaled by sleep-associated surges in LH.
   Answer: True

7. Sperm are moved from the seminiferous tubules directly to the epididymis.
   Answer: False

8. Semen has a slightly alkaline pH.
   Answer: True

9. The stroma of the ovaries is composed of simple cuboidal epithelium.
   Answer: False

10. During early fetal development, germ cells in the ovaries differentiate into oogonia.
    Answer: True

11. Fertilization may occur up to 24 hours after ovulation.
    Answer: True

12. The uterus is supported and held in position by several skeletal muscles.
    Answer: False

13. The mucosa of the vagina has an acid environment.
    Answer: True

14. The region between the labia minora is called the hymen.
    Answer: False

15. The secretion of milk from the mammary glands following delivery is largely due to the hormone prolactin.
    Answer: True

## Short Answer

*Write the word or phrase that best completes each statement or answers the question.*

1. The branch of medicine that deals with the diagnosis and treatment of diseases of the female reproductive system is _____.
   Answer: gynecology

2. The pouch that supports the testes is the _____.
   Answer: scrotum

3. The lobules of the testis contain _____ _____.
   Answer: seminiferous tubules

4. The portion of a sperm that contains an enzyme necessary for the penetration of the secondary oocyte is the _____.
   Answer: acrosome

5. The period of time when secondary sex characteristics begin to develop is referred to as _____.
   Answer: puberty

6. A supportive structure of the male reproductive system that carries blood vessels, autonomic nerves, and lymphatic vessels is the ____.
   Answer: spermatic cord

7. A method of sterilization in males where a portion of each ductus deferens is removed is called
   Answer: vasectomy

8. The two dorsal masses of erectile tissue of the penis are the ____.
   Answer: corpora cavernosa penis

9. The mass of erectile tissue surrounding the male urethra is the ____.
   Answer: corpus spongiosum penis

10. The distal end of the corpus spongiosum is a slightly enlarged region called the ____.
    Answer: glans penis

11. A surgical procedure in which part or all of the prepuce is removed is called ____.
    Answer: circumcision

12. The erection of the penis is due to a _____ reflex.
    Answer: parasympathetic

13. The remnants of an ovulated mature follicle form the ____.
    Answer: corpus luteum

14. The degenerated corpus luteum is called ____.
    Answer: corpus albicans

15. A growing follicle forms a clear layer of glycoprotein called the ____.
    Answer: zona pellucida

16. The broad ligament that forms part of the outer layer of the uterus is the ____.
    Answer: perimetrium

17. The middle muscular layer of the uterus is called ____.
    Answer: myometrium

18. The innermost part of the uterine wall is the ____.
    Answer: endometrium

19. The structure that serves as a passageway for menstrual flow and childbirth is the ____.
    Answer: vagina

20. The region between the vagina and the anus is known as the ____.
    Answer: clinical perineum

21. The mammary glands are modified ____ glands.
    Answer: sudoriferous (sweat)

22. The circular pigmented area of skin surrounding the nipple is called the ____.
    Answer: areola

23. The external genitalia of the female are collectively called _____.
    Answer: vulva

24. The hormone released by the hypothalamus that effects both the female and male reproductive system is _____.
    Answer: GnRh

25. The human sexual response has ___ stages.
    Answer: 4

## Matching

*Choose the item from Column 2 that best matches each item in Column 1.*

1.  Column 1: Testes do not descend.
    Column 2: cryptorchidsm

2.  Column 1: The onset of menstruation
    Column 2: menarche

3.  Column 1: The absence of menstruation.
    Column 2: amenorrhea

4.  Column 1: painful menstruation due to forceful contractions of the uterus.
    Column 2: dysmenorrhea

5.  Column 1: Presence of both male and female sex organs in one individual.
    Column 2: hermaphroditism

6.  Column 1: A displaced urethral opening.
    Column 2: hypospadias

## Essay

*Write your answer in the space provided or on a separate sheet of paper.*

1.  Describe the path of sperm through the male reproductive system.
    Answer: Sperm is produced in the seminiferous tubules of the testis, moved to the straight tubules to the rete testis. The sperm are then transported into the epididymis to the ductus deferens, where the sperm can be stored. During ejaculation, sperm is moved into the ejaculatory duct and finally into the urethra, and the terminal duct of the system.

2.  Name and explain the function of the accessory sex glands of the male reproductive system.
    Answer: The accessory sex glands secrete most of the liquid portion of the semen. The paired seminal vesicles secrete an alkaline, viscous fluid that contains fructose, prostaglandins, and clotting proteins. The alkalinity of the fluid helps neutralize the acid in the female reproductive tract.
    The prostate gland secretes a milky, slightly acidic fluid that contains nutrients and several enzymes.
    The paired bulbourethral glands secrete an alkaline substance that protects sperm by neutralizing the acid environment of the urethra, and a mucus that lubricates the end of the penis during intercourse.

3.  Define the female reproductive cycle.
    Answer: The female reproductive cycle refers to the menstrual and ovarian cycle. The menstrual cycle is the series of monthly changes in the non-pregnant female, the ovarian cycle describes the monthly changes occurring in the ovaries.

4. Name the sexually transmitted diseases caused by viruses.
   Answer: AIDS, genital herpes, genital warts

## CHAPTER 24   Development and Inheritance

## Multiple-Choice

*Choose the one alternative that best completes the statement or answers the question.*

1. Of the sperm cells introduced into the vagina, how many actually reach the secondary oocyte?
   A)  12 percent
   B)  8 percent
   C)  5 percent
   D)  2 percent
   E)  less than 1 percent
   Answer: E

2. A secondary oocyte is viable for approximately _____ hours after ovulation.
   A)  6
   B)  12
   C)  24
   D)  36
   E)  48
   Answer: C

3. Embryology refers to the study of development from the fertilized egg through the _____ week.
   A)  second
   B)  fifth
   C)  eighth
   D)  ninth
   E)  tenth
   Answer: C

4. The term neonatal period defines the first _____ days after birth.
   A)  50
   B)  42
   C)  36
   D)  12
   E)  8
   Answer: B

5. Birth in general occurs _____ weeks after fertilization.
   A)  30
   B)  32
   C)  34
   D)  36
   E)  38
   Answer: E

6. Before a sperm cell can penetrate an ovum it has to undergo a series of changes called _____.
   A)  capacitation.
   B)  lubrication.
   C)  pulpation.
   D)  dislocation.
   E)  maturation.
   Answer: A

7. For fertilization to occur the sperm has to penetrate layers surrounding the oocyte. The first layer is the
   A) zona glomerulosa.
   B) zona pellucida.
   C) corona radiata.
   D) corona granulosa.
   E) reticular layer.
   Answer: C

8. Fraternal twins
   A) develop from a single fertilized ovum.
   B) are the same age but genetically different.
   C) are genetically identical.
   D) are always the same sex.
   E) are fertilized by one sperm cell.
   Answer: B

9. The rapid mitotic cell division of the zygote that occurs immediately after fertilization is called
   A) development.
   B) differentiation.
   C) gravidation.
   D) cleavage.
   E) cytokinesis.
   Answer: D

10. Successive cleavages produce a solid mass of tiny cells called
    A) embryo.
    B) blastocyst.
    C) morula.
    D) zygote.
    E) trophoblast.
    Answer: C

11. The chorion and placenta are formed from the
    A) trophoblast.
    B) inner cell mass.
    C) trophoblast and inner cell mass.
    D) embryo.
    E) morula.
    Answer: C

12. The process of implantation occurs approximately ____ days after fertilization.
    A) 6
    B) 8
    C) 10
    D) 12
    E) 14
    Answer: A

13. Blood levels of hCG reach a maximum during the
    A) third week of pregnancy.
    B) fifth week of pregnancy.
    C) ninth week of pregnancy.
    D) twelfth week of pregnancy.
    E) eighteenth week of pregnancy.
    Answer: C

14. During vitro fertilization,
    A) a woman is artificially inseminated with the husband's semen.
    B) the husband's semen is used to artificially inseminate a secondary oocyte donor.
    C) fertilization happens in a laboratory dish.
    D) a woman is given hormone treatment and is then artificially inseminated with donated semen.
    E) All of the above.
    Answer: C

15. The term fetal period refers to the time from the ___ month until birth.
    A) 1$^{st}$
    B) 2$^{nd}$
    C) 3$^{rd}$
    D) 4$^{th}$
    E) 5$^{th}$
    Answer: B

16. The placenta is functioning by the end of the
    A) first month.
    B) second month.
    C) third month.
    D) fourth month.
    E) immediately after implantation.
    Answer: C

17. Which of the following develops last?
    A) the ectoderm
    B) the embryonic disc
    C) the yolk sac
    D) the mesoderm
    E) the endoderm
    Answer: D

18. The ectoderm, mesoderm, and endoderm are attached to the trophoblast by the
    A) embryonic disc.
    B) embryonic membranes.
    C) chorion.
    D) body stalk.
    E) chorionic membranes.
    Answer: D

19. Which of the following embryonic membranes becomes the principal part of the embryonic part of the placenta?
    A) endoderm
    B) chorion
    C) mesenchyme
    D) allantois
    E) amnion
    Answer: B

20. Which of the following serves as an early site of blood formation?
    A) yolk sac
    B) allantois
    C) chorion
    D) amnion
    E) amniotic fluid
    Answer: B

21. Wastes leave the fetus through the
    A) umbilical veins.
    B) intervillous spaces.
    C) umbilical arteries.
    D) chorionic villi.
    E) ductus venous.
    Answer: C

22. The umbilical vein goes to the
    A) heart of the fetus.
    B) liver of the fetus.
    C) lungs of the fetus.
    D) vena cava of the fetus.
    E) kidney of the fetus.
    Answer: B

23. The only fetal vessel that carries fully oxygenated blood is the
    A) umbilical vein.
    B) umbilical artery.
    C) ductus arteriosus.
    D) pulmonary trunk.
    E) common iliac arteries.
    Answer: A

24. How long is the corpus luteum maintained after fertilization?
    A) two weeks
    B) six weeks
    C) at least two months
    D) at least three to four months
    E) no longer than six months
    Answer: D

25. Which of the following develop from the ectoderm?
    A) lens and cornea
    B) nervous system
    C) epidermis
    D) All of the above.
    E) None of the above.
    Answer: D

26. Which of the following develop from the mesoderm?
    A) nervous system
    B) cardiac muscle
    C) epidermis
    D) All of the above.
    E) None of the above.
    Answer: B

27. The hormone believed to stimulate the development of breast tissue for lactation is
    A) estrogen.
    B) progesterone.
    C) relaxin.
    D) hCG.
    E) hCS.
    Answer: E

28. All of the following are pregnancy induced changes EXCEPT
   A) increased cardiac output.
   B) decreased gastrointestinal tract motility.
   C) increased tidal volume.
   D) decrease in respiratory reserve.
   E) all of the above are pregnancy induced changes.
   Answer: E

29. Pharmaceutical companies use human placentas as a source for
   A) hormones.
   B) drugs.
   C) blood.
   D) None of the above.
   E) All of the above.
   Answer: E

30. All of the following can be detected by amniocentesis EXCEPT
   A) Down's Syndrome.
   B) diabetes.
   C) hemophilia.
   D) sickle-cell anemia.
   E) cystic fibrosis.
   Answer: B

31. In which month of fetal development is subcutaneous fat deposited?
   A) 3$^{rd}$
   B) 4$^{th}$
   C) 6$^{th}$
   D) 8$^{th}$
   E) 9$^{th}$
   Answer: D

32. The heartbeat can first be detected in the _____ month of pregnancy.
   A) 1$^{st}$
   B) 2$^{nd}$
   C) 3$^{rd}$
   D) 4$^{th}$
   E) 5$^{th}$
   Answer: C

33. The hormone that stimulates uterine contractions is
   A) vasopressin.
   B) oxytocin.
   C) prolactin.
   D) relaxin.
   E) estrogen.
   Answer: B

34. The major hormone promoting lactation is
   A) oxytocin.
   B) vasopressin.
   C) prolactin.
   D) estrogen.
   E) progesterone.
   Answer: C

35. Milk ejection is promoted by
    A) oxytocin.
    B) vasopressin.
    C) prolactin.
    D) estrogen.
    E) progesterone.
    Answer: A

36. The hormones found in the birth control pill are
    A) estrogen and LH.
    B) LH and FSH.
    C) progesterone and FSH.
    D) estrogen and progesterone.
    E) estrogen and inhibin.
    Answer: D

37. Genes that control the same inherited trait that occupy the same position on homologous chromosomes are called
    A) carries.
    B) alleles.
    C) dominant.
    D) recessive.
    E) equal.
    Answer: B

38. Which of the following is caused by a dominant gene?
    A) sickle cell anemia
    B) fetal alcohol syndrome
    C) Down's syndrome
    D) Huntington's disease
    E) phenylketonuria
    Answer: D

39. The nuclei of all human cells, except gametes, contain
    A) 46 pairs of chromosomes.
    B) 36 pairs of chromosomes.
    C) 28 pairs of chromosomes.
    D) 23 pairs of chromosomes.
    E) 22 pairs of chromosomes.
    Answer: D

40. The actual genetic makeup is contained in the
    A) phenotype.
    B) genotype.
    C) Punnett square.
    D) dominant genes.
    E) All of the above.
    Answer: B

41. Which of the following is a teratogen?
    A) cocaine
    B) alcohol
    C) nicotine
    D) pesticides
    E) all of the above
    Answer: E

42. Which of the following is a recessive trait?
    A) freckles
    B) baldness
    C) widow's peak
    D) large eyes
    E) normal hearing
    Answer: B

43. Which of these diseases is a recessive condition?
    A) cystic fibrosis
    B) diabetes insipidus
    C) Huntington's disease
    D) Both A and B.
    E) All of the above.
    Answer: A

44. Which of the following is a condition which is inherited by sex-linked traits?
    A) Huntington's disease
    B) diabetes insipidus
    C) cystic fibrosis
    D) hemophilia
    E) PKU
    Answer: D

45. Klinefelter's syndrome is a sex chromosome disorder with the following genotype:
    A) XYY
    B) XXY
    C) XXX
    D) XO
    E) XY
    Answer: B


## True-False

*Write T if the statement is true and F if the statement is false.*

1. Monozygotic twins develop from a single fertilized ovum.
   Answer: True

2. The zona pellucida is a layer of clear glycoprotein.
   Answer: True

3. The months of development after the second month are considered the fetal period.
   Answer: True

4. The allantois becomes the principal embryonic part of the placenta.
   Answer: False

5. The opening in the septum of a fetus between the right and the left atrium is called the foramen magnum.
   Answer: False

6. The placenta and the corpus luteum produce the hormone relaxin.
   Answer: True

7. Chorionic villi sampling is a test that picks up more defects than amniocentesis.
   Answer: False

8. The principal stimulus in maintaining prolactin secretion during lactation is the sucking action of the infant.
   Answer: True

9. Following birth, prolactin levels start increasing.
   Answer: False

10. Coitus interruptus refers to withdrawal of the penis from the vagina just prior to ejaculation.
    Answer: True

11. Homologous chromosomes contain similar traits.
    Answer: True

12. The sex chromosomes are also responsible for the transmission of a number of nonsexual traits.
    Answer: True

13. Maternal smoking has no impact on the developing fetus.
    Answer: False

14. X and Y chromosomes are sex chromosomes.
    Answer: True

15. PKU is a sex linked inherited disorder.
    Answer: False


## Short Answer

*Write the word or phrase that best completes each statement or answers the question.*

1. Twins that develop from a single fertilized ovum are called ____.
   Answer: identical

2. When the morula forms a hollow ball of cells, it is then called ____.
   Answer: blastocyst

3. The outer layer of a blastocyst is called the ____.
   Answer: trophoblast

4. If an embryo implants in a location outside the uterine cavity it is called an _____ pregnancy.
   Answer: ectopic

5. The three primary germ layers are the ectoderm, endoderm, and _____.
   Answer: mesoderm

6. All nervous tissue develops from the _____.
   Answer: ectoderm

7. A procedure in which a husband's semen is used to artificially inseminate a fertile secondary oocyte donor, followed by a transfer of the morula to the infertile wife is called ____.
   Answer: embryo transfer

8. The ectoderm separates to form a fluid filled space called the ____.
   Answer: amniotic cavity

9. During embryonic life, finger-like projections grow into the endometrium of the uterus. These projections are called ____.
   Answer: chorionic villi

10. After the birth of the baby, the placenta detaches from the uterus and is termed the _____.
    Answer: afterbirth

11. Blood passes from the fetus to the placenta via two _____ in the umbilical cord.
    Answer: umbilical arteries

12. The period of time a zygote, embryo, and fetus are carried in the female reproductive tract is called _____.
    Answer: gestation

13. The process by which uterine contractions expel the fetus through the vagina to the outside is called _____.
    Answer: labor

14. The term emesis gravidarum means _____ _____.
    Answer: morning sickness

15. The secretion and ejection of milk by the mammary glands is called _____.
    Answer: lactation

16. During the first few days after birth, the mammary glands secrete a fluid called _____.
    Answer: colostrum

17. Sterilization of females is usually achieved by _____.
    Answer: tubal ligation

18. The branch of biology that studies inheritance is _____.
    Answer: genetics

19. The way the genetic makeup is expressed in the body is referred to as _____.
    Answer: phenotype

20. Inherited traits that are not controlled by one gene, but by the combined effect of many, are referred to as _____.
    Answer: polygenic inheritance

21. Intrauterine exposure to alcohol results in _____.
    Answer: fetal alcohol syndrome

22. The chromosomes, which are not sex chromosomes, are called _____.
    Answer: autosomes

23. A permanent change in a gene that might be inheritable is a _____.
    Answer: mutation

24. An abortion that occurs without apparent cause is a _____.
    Answer: spontaneous abortion

25. Down's syndrome is a disorder involving chromosome ___.
    Answer: 21


**Matching**

*Choose the time form Column 2 that best matches each item in Column 1.*

1. Column 1: tubal ligation
   Column 2: sterilization

2. Column 1: vasectomy
   Column 2: sterilization

3. Column 1: oral contraception
   Column 2: hormonal method

4. Column 1: a small object of plastic, copper, or stainless steel inserted into the uterine cavity.
   Column 2: intrauterine device

5. Column 1: condoms
   Column 2: barrier method

6. Column 1: vaginal pouch
   Column 2: barrier method

7. Column 1: cervical cap
   Column 2: barrier method

8. Column 1: diaphragm
   Column 2: barrier method

9. Column 1: spermicidal agents
   Column 2: chemical method

10. Column 1: rhythm method
    Column 2: natural method

11. Column 1: sympto-thermal method
    Column 2: natural method

## Essay

*Write your answer in the space provided or on a separate sheet of paper.*

1. Name the primary germ layers and the main structures developing from each
   Answer:  1. Endoderm: it becomes the epithelial lining of the gastrointestinal tract, respiratory tract, and a number of other organs.
   2. Mesoderm: it forms the peritoneum, muscle, bone, and other connective tissue.
   3. Ectoderm: develops into the skin and nervous system.

2. Describe the stages of labor.
   Answer: 1.  The stage of dilation is the time from the onset of labor to the complete dilation of the cervix.
   2. The stage of expulsion is the time from complete cervical dilation to delivery.
   3. The placental stage is the time after delivery until the placenta is expelled.

3. Name the principal methods of birth control, and give an example of each.
   Answer: sterilization- vasectomy and tubal ligation
   Hormonal methods- oral contraceptive
   Intrauterine devices- copper T
   Barrier methods- condoms
   Chemical methods- spermicides
   Natural methods- rhythm, coitus interruptus, induced abortion